Test Automation Engineering Handbook

Learn and implement techniques for building robust test automation frameworks

Manikandan Sambamurthy

‹packt›

BIRMINGHAM—MUMBAI

Test Automation Engineering Handbook

Copyright © 2023 Packt Publishing

Group Product Manager: Gebin George

Publishing Product Manager: Kunal Sawant

Senior Editor: Rounak Kulkarni

Technical Editor: Jubit Pincy

Copy Editor: Safis Editing

Project Coordinator: Manisha Singh

Proofreader: Safis Editing

Indexer: Subalakshmi Govindhan

Production Designer: Vijay Kamble

Developer Relations Marketing Executive: Sonakshi Bubbar

Business Development Executive: Debadrita Chatterjee

First published: January 2023

Production reference: 1221222

Published by Packt Publishing Ltd.

Livery Place

35 Livery Street

Birmingham

B3 2PB, UK.

ISBN 978-1-80461-549-2

www.packt.com

To my late mother, Gomathy, whose unconditional love inspires me to always do my best. To my dad, wife, children, and sisters, whose support gives me the confidence to complete this work.

– Manikandan Sambamurthy

Contributors

About the author

Manikandan Sambamurthy has been a software engineer, particularly focused on quality engineering, for over 15 years. He has helped several enterprise organizations formulate and implement their test strategies successfully. He has built and mentored diverse and smooth-functioning quality engineering teams throughout his career. Manikandan is a prolific tester and specializes in full stack test automation. He is a lifelong learner and possesses tremendous knowledge of both frontend and backend test technologies. He has led various continuous improvement efforts to improve the quality and productivity of entire software engineering teams. With his deep understanding and expertise in the quality domain, he has saved considerable costs across organizations through meticulous customizations of test automation frameworks.

I would like to thank my wife and two children, who have supported me throughout the process of writing this book. Sincere thanks to the tremendously helpful team at Packt for getting this book to completion. I would also like to thank the technical reviewers for all their valuable feedback.

About the reviewers

Ayesha Janvekar is an IT professional based in San Diego, California. She is currently working as a software development manager with more than 9 years of hands-on QA experience. Ayesha is an experienced IT leader and is passionate about software testing. She has a bachelor's degree in electronics and communication engineering, an MBA in marketing and international business, and a master's in information technology and management from Chicago. She believes in translating her work into actionable insights. Having strong project management skills and business acumen, Ayesha helps deliver pragmatic technical solutions.

Meir Blachman, an Israeli Jew, completed a bachelor's degree in computer science after high school. In 2014, he joined the Israeli army and served in the 8200 unit for 5.5 years. Meir started his service as an automation engineer and introducing teams to web automation, and transitioned to backend engineering focusing on CI/CD tooling and automation around the development team components. In 2019, he started working for Microsoft on the Cloud App Security group, specifically on the Conditional Access App Control product. During this time, Meir learned how the web works, and specifically browsers, involving the HTML, CSS, and JavaScript standards and how browsers process this content. He has also invented multiple patents through Microsoft.

Chris Wallander is the principal SDET and test automation architect at TaxAct and has over 10 years of experience in test automation and quality assurance, as well as a solid foundation in DevOps and Agile methodologies. Trained and certified through the International Software Testing Qualification Board, his focus is on analyzing enterprise systems, and then building test automation frameworks to support them. Outside of work, he enjoys developing and testing video games, AI/ML development, building robots, and smithing medieval armor.

Table of Contents

Preface   xiii

Part 1: The Basics

1

Introduction to Test Automation   3

Getting familiar with software testing	3	Test automation challenges	12
Knowing the importance of testing	4	Finding and handling regression bugs	13
Tasks involved in testing	4	Test automation metrics	13
Testing in an world	5	**Exploring the roles in quality**	
Defect management in testing	6	**engineering**	**15**
Shift-Right and Shift-Left propositions	7	Test automation engineer	15
Quality and DevOps	7	SDET	16
Challenges in testing	8	**Familiarizing yourself with common**	
Test early, test often	8	**terminologies and definitions**	**17**
Understanding test automation	**9**	**Summary**	**19**
Agile test automation	11	**Questions**	**20**

2

Test Automation Strategy   21

Technical requirements	**21**	Test automation environment	24
Knowing your test		Implementing an Agile	
automation strategy	**22**	test automation strategy	26
Test automation objectives	22	Reporting the test results	28
Gathering management support	23	**Devising a good test**	
Defining the scope of test automation	24	**automation strategy**	**29**

Selecting the right tools and training   29
Standards of the test automation framework   30
Testing in the cloud   34

Understanding the test pyramid   **35**
Unit/component tests   35
Integration/API tests   36
E2E/System/UI tests   36

Structuring the test cycles   37

**Familiarizing ourselves with
common design patterns**   **37**
Using design patterns in test automation   37

Summary   **42**
Questions   **42**

3

Common Tools and Frameworks   43

Technical requirements   **43**
**The basic tools for every
automation engineer**   **43**
The CLI   44
Git   49

**Common test automation
frameworks**   **56**
Selenium   57
Appium   60
Cypress   62

JMeter   63
AXE   65

Choosing the right tool/framework   **66**
Selecting a performance testing tool   66
Selecting an API testing tool   67
Selecting a web testing tool   67
Mobile   68
Common considerations   68

Summary   **69**
Questions   **69**

Part 2: Practical Affairs

4

Getting Started with the Basics   73

Technical requirements   **73**
Getting more familiar with Git   **74**
Committing a change   74
Resolving merge conflicts   77
Additional Git commands   80

Using an IDE   **81**
Choosing an IDE   81

Downloading and setting up VS Code   82

Introduction to JavaScript   **84**
Why learn JavaScript?   84
Running a JavaScript program   84
JavaScript basics   85

Summary   **101**
Questions   **101**

5

Test Automation for Web 103

Technical requirements 103
Why Cypress? 104
Installing and setting up Cypress 104
Creating your first test in Cypress 109
Creating arrow functions in JavaScript 109
Creating callback functions in JavaScript 110
Writing our first spec 112
Becoming familiar with the spec structure 112
Executing our first spec 113

Employing selectors and assertions 116

Working with selectors 116
Asserting on selectors 117
Intercepting API calls 119
Additional configurations 120
Considerations for web automation 121
Limitations of Cypress 121
Web automation considerations 122

Summary 122
Questions 123

6

Test Automation for Mobile 125

Technical requirements 125
Getting to know Appium 126
What is Appium? 126
Advantages of using Appium 126

Knowing WebdriverIO and
its advantages 127

Setting up Appium and
WebdriverIO 127
Appium installation 127
Configuring an Android emulator 129
Configuring WebdriverIO with Appium 132
WebdriverIO Android configuration 134
Appium Inspector installation
and configuration 136

Writing our first mobile test 138
JavaScript functions with async/await 138
First Appium/WebdriverIO test 140

Key considerations for mobile
automation 141
Areas of complexity 141
iOS-specific considerations 142

Optimizing our mobile
automation framework 144
Summary 146
Questions 146

7

Test Automation for APIs   147

Technical requirements	148	Using snippets for asserting an API response	162
Getting started with Postman	148	Understanding Postman variables	164
Basics of REST API testing	148	Chaining API requests	166
Downloading the Postman application	149	Various ways to execute tests	167
Creating and managing workspaces	150	**Key considerations for API**	
Sending GET and POST requests	**151**	**automation**	**171**
Making a GET API request	152	Effective API test automation	171
Making a POST API request	155	Testing GraphQL versus REST APIs	172
Organizing API requests using collections	157	**Summary**	**172**
Writing automated API tests	**162**	**Questions**	**173**

8

Test Automation for Performance   175

Technical requirements	175	Java essentials for JMeter	190
Getting started with JMeter	175	A quick introduction to Java	191
What is JMeter and how does it work?	176	Using the JSR233 assertion	193
Installing JMeter	176	**Considerations for**	
Automating a performance test	**179**	**performance testing**	**195**
Building and running our first		**Summary**	**196**
performance test	179	**Questions**	**196**
Working with assertions	184		
Working with tests via the command line	186		
Using the HTTP(S) Test Script Recorder	189		

Part 3: Continuous Learning

9

CI/CD and Test Automation   199

Technical requirements	199	Unit/component tests	205
What is CI/CD?	199	API tests	205
CI/CD process	200	E2E tests (API and UI)	205
CI basics	201	Smoke tests	206
CD and deployment pipeline	202	Addressing test failures	207
Test automation strategies for CI/CD	204	GitHub Actions CI/CD	208
		Summary	212
		Questions	212

10

Common Issues and Pitfalls   213

Recurrent issues in test automation	213	Not taking a lean approach	216
Unrealistic expectations of automated testing	213	Not having a plan for test data needs	217
Inadequate manual testing	214	Test automation anti-patterns	217
Not focusing on automating the right things	214	Coding and design anti-patterns in test automation	217
A lack of understanding of the system under test	215	Process-oriented anti-patterns in test automation	223
Overlooking test maintenance	215		
Not choosing the right tools	215	Summary	226
Under-investing in test environments	216	Questions	227
Taking a siloed approach	216		

Appendix A: Mocking API Calls   229

How API mocking works	229	Considerations for API mocking	233
Mocking API calls using Postman	230	Summary	234

Assessments   235

Chapter 1, Introduction
to Test Automation   235

Chapter 2, Test Automation Strategy   235

Chapter 3, Common Tools
and Frameworks   236

Chapter 4, Getting Started with
the Basics   237

Chapter 5, Test Automation for Web   237

Chapter 6, Test Automation
for Mobile   238

Chapter 7, Test Automation for APIs   238

Chapter 8, Test Automation
for Performance   239

Chapter 9, CI/CD and Test
Automation   239

Chapter 10, Common Issues
and Pitfalls   240

Index   243

Other Books You May Enjoy   254

Preface

Quality is indispensable in every aspect of our lives, and software systems are no different. Throughout my career in software quality (spanning more than 15 years), I have constantly perceived the increasing appetite for achieving greater quality as part of the software development processes. As software applications grow more complex, the quality processes and tools around them have to keep up with all the intricacies. Finding problems in their products early helps organizations move faster and contributes to positive business outcomes.

Test automation is at the core of most of the quality endeavors within the software development life cycle. Test automation engineering is the process of enhancing the quality of software applications through automated tests. If test automation practices are correctly baked into the product development life cycle from the commencement of a project, it enables an invaluable feedback loop resulting in tremendous quality improvements. Test automation also reduces testing time, which allows resources to be used on other worthwhile efforts. Automation-first is the new norm across all software testing activities.

Testing is not a standalone activity and is viewed as an integral part of the software development process. Embedded quality teams have become the new normal, and hence communication and collaboration are key to the success of any test automation undertaking. There has been a paradigm shift in the way software developers view test automation and in how testers regard their own roles.

I will introduce you to various aspects of test automation in this book through simple definitions and practical examples. I will also review the different quality roles in the market and the skills it takes to be successful in each. The testing world is evolving at a great speed, and this book will help you solidify your fundamentals in test automation so that you can keep up with further progress in the field. Due to the cross-functional nature of the quality engineering discipline, it is necessary now more than ever to bridge the gap in test automation skills across software engineering and DevOps disciplines. This book will act as an excellent ally in filling this gap by helping you learn the multiple facets of test automation. I will provide relevant examples and guide you through various commonly used test automation frameworks in the industry.

Who this book is for

This will book will help anyone who wishes to enter the realm of software quality and test automation. It lays out the necessary fundamentals first and builds on them gradually. Experienced quality engineers, software engineers, and manual testers can gain practical insights throughout this book. Let's see how this book helps each role:

- **Quality engineers**: Quality engineers who have a test automation foundation but are seeking to upskill in test automation across various platforms will learn relevant knowledge. They will also get a deeper grasp of test automation strategies and considerations. It can also act as a supplement for quality engineering interviews.

- **Manual testers**: Manual testers who wish to switch over to test automation engineering roles can use this book as the foundation and guide for their journey.

- **Software engineers**: Software engineers who would like to get quickly up and running with a test automation framework can use this book as a guide. It will not only help familiarize with a framework but also review the additional considerations needed when adopting it.

What this book covers

Chapter 1, Introduction to Test Automation, introduces readers to the world of testing and test automation as a software engineering practice. It also provides insights into the roles in quality engineering and familiarizes readers with common terms and definitions.

Chapter 2, Test Automation Strategy, discusses test automation strategies and guides readers on how to devise one. It also explores the test automation pyramid and introduces readers to common test automation design patterns.

Chapter 3, Common Tools and Frameworks, establishes the basic tools that are used in test automation. It also explores in detail some of the commonly used test automation frameworks and helps you choose the right framework.

Chapter 4, Getting Started with the Basics, covers some advanced Git commands and introduces readers to IDEs. Readers are then taken through a crash course on JavaScript.

Chapter 5, Test Automation for Web, explores the tool **Cypress** in detail. It covers the unique aspects of Cypress and goes over the necessary steps of installation and setup. It also helps readers to write a basic test and employ assertions in it. It explains how to intercept API calls and reviews additional considerations for web test automation.

Chapter 6, Test Automation for Mobile, introduces readers to the mobile test automation framework Appium. It goes over the installation and setup and enables readers to write their first mobile automation test using Appium. It also deals with some key considerations of mobile test automation.

Chapter 7, Test Automation for API, explores in detail how to use the Postman tool to perform RESTful API testing. It also provides readers with options to automate API tests in Postman.

Chapter 8, *Test Automation for Performance*, introduces readers to the JMeter tool and explains how to create and execute performance tests using it. It explores the various useful aspects of JMeter for building a complete performance test. It also reviews certain key considerations when designing performance tests.

Chapter 9, *CI/CD and Test Automation*, establishes the concepts of CI/CD and deals with how test automation strategies apply to the CI/CD methodology. It also takes readers through a demo of GitHub Actions, which is a built-in CI/CD tool in GitHub.

Chapter 10, *Common Issues and Pitfalls*, educates readers on some of the common issues faced in test automation and recommends a solution for each of them. It also provides tips on avoiding any pitfalls when implementing test automation projects.

Appendix A, *How to Dockerize Automated Tests*, helps readers to use container technology to build and run automated tests.

To get the most out of this book

Although this book aims to build fundamental test automation knowledge, it is good to have a basic knowledge of software testing terminology. It will also be helpful to know the basics of the software development life cycle and agile software practices.

Software/hardware covered in the book	Operating system requirements
Visual Studio Code (VC Code)	Windows, macOS, or Linux
Python 3.5+	Chrome, Firefox, Edge
Java Runtime Environment (JRE) 1.8+	
Terminal (macOS) / PowerShell (Windows)	
Cypress version 11.2.0	
Node.js	
Java Development Kit (JDK)	
Android Studio	
Postman (version 9.3.15)	
Newman command-line tool	
JDK 8 and JRE 8 or higher	

Conventions used

There are a number of text conventions used throughout this book.

`Code in text`: Indicates code in action, commands, keywords, folder names, filenames, file extensions, and pathnames.

Here is an example: "The `setTimeout()` function calls a method after a specified wait in milliseconds. For example, `setTimeout(() => console.log('hello!'), 5000)` prints the message after a wait of 5 seconds."

A block of code is set as follows:

```
describe ("First Android Spec", () => {
it ("to find element by accessibility id", async () => {
const animationOption = await $("~Animation");
```

Any command-line input or output is written as follows:

```
appium driver install xcuitest
appium driver install uiautomator2
```

Bold: Indicates a new term, an important word, or words that you see onscreen. Here is an example: "Use the **Settings** option in the **Preferences** menu for additional configuration."

> **Tips or important notes**
> Appear like this.

Download the example code files

You can download the example code files for this book from GitHub at `https://packt.link/Uhjqi`. If there is an update to the code, it will be updated in the GitHub repository.

We also have other code bundles from our rich catalog of books and videos available at `https://github.com/PacktPublishing/`. Check them out!

Get in touch

Feedback from our readers is always welcome.

General feedback: If you have questions about any aspect of this book, email us at `customercare@packtpub.com` and mention the book title in the subject of your message.

Errata: Although we have taken every care to ensure the accuracy of our content, mistakes do happen. If you have found a mistake in this book, we would be grateful if you would report this to us. Please visit www.packtpub.com/support/errata and fill in the form.

Piracy: If you come across any illegal copies of our works in any form on the internet, we would be grateful if you would provide us with the location address or website name. Please contact us at copyright@packt.com with a link to the material.

If you are interested in becoming an author: If there is a topic that you have expertise in and you are interested in either writing or contributing to a book, please visit authors.packtpub.com.

Share Your Thoughts

Once you've read *Test Automation Engineering Handbook*, we'd love to hear your thoughts! Scan the QR code below to go straight to the Amazon review page for this book and share your feedback.

https://packt.link/r/1804615498

Your review is important to us and the tech community and will help us make sure we're delivering excellent quality content.

Download a free PDF copy of this book

Thanks for purchasing this book!

Do you like to read on the go but are unable to carry your print books everywhere?

Is your eBook purchase not compatible with the device of your choice?

Don't worry, now with every Packt book you get a DRM-free PDF version of that book at no cost.

Read anywhere, any place, on any device. Search, copy, and paste code from your favorite technical books directly into your application.

The perks don't stop there, you can get exclusive access to discounts, newsletters, and great free content in your inbox daily

Follow these simple steps to get the benefits:

1. Scan the QR code or visit the link below

https://packt.link/free-ebook/9781804615492

2. Submit your proof of purchase
3. That's it! We'll send your free PDF and other benefits to your email directly

Part 1: The Basics

In this part, we will begin with the basics of testing and test automation. You will learn about the considerations that go into devising effective test automation strategies. You will also get familiarized with Git and grasp the fundamentals of prominent test automation frameworks. By the end of this section, you will recognize and understand commonly used test automation terms, frameworks, and tools.

This part has the following chapters:

- *Chapter 1, Introduction to Test Automation*
- *Chapter 2, Test Automation Strategy*
- *Chapter 3, Common Tools and Frameworks*

1

Introduction to Test Automation

One of the indisputable things when building and delivering any product is its quality. End users of a product will not settle for anything less than superior when it comes to product quality. In this chapter, we will dive deep into how testing and test automation help achieve this level of quality in a software product. The first few pages will introduce the reader to testing and test automation. Later in the chapter, we will dive deeper into the subject of test automation.

Quality is everyone's responsibility in a team, and this chapter provides various practical pointers to help establish that collaboration. Additionally, we will see how automated tests add another layer of complexity to a project and help understand the best practices to accomplish coherent test automation. We will also look at the lineup of development and test automation processes to provide a reliable and bug-free product experience for customers.

Here are the key topics that you will learn by the end of this chapter:

- Getting familiar with software testing
- Introducing test automation
- Exploring the roles in quality engineering
- Familiarizing yourself with common terminologies and definitions

Getting familiar with software testing

Testing as an activity is critical to delivering a trustworthy product and forms a strong foundation for building reliable test automation. So, being a good tester makes you a more effective test automation engineer.

Software testing is an indispensable task in any software development project that is mainly done with the goals of validating the specified product requirements and finding bugs. These bugs can be functional or non-functional in nature. Functional bugs include deviations from the specified requirements or product specifications. Usually, non-functional issues are performance-based or security-based. The primary goals of testing are usually interwoven at multiple levels but can be broken down at a high level as follows:

- **Functional**: Checking the business functionalities of the software system:

    - Compliance: Regulator agencies, government agencies, and more

    - Portability: Cross-browser testing, mobile support, and more

    - Usability: Support for disabled users

    - Maintainability: Vendor support

- **Non-functional**: Checking the non-functional aspects of the software system, which are never tackled by functional testing:

    - Reliability: Operational up time, failovers, business continuity, and more

    - Security: Vulnerability, penetration testing, and more

    - Performance: Load, stress testing, and more

Now that we have seen a quick introduction to testing, let us understand why testing is so critical.

Knowing the importance of testing

Even though software's quality is initially laid out by the design and architecture, testing is the core activity that gives stakeholders much-needed confidence in the product. By verifying the behavior of the product against documented test cases, the activity of testing helps uncover bugs and address other design issues promptly. By preventing and identifying bugs early in the software development life cycle, testing helps both the engineering and business teams to increase customer satisfaction and reduce overall operating costs. The valuable insights that the team derives from the repeated testing of the product can be further used to improve the efficiency of the software development process. Therefore, it is imperative to conduct testing for achieving the established goals of a successful product launch.

Tasks involved in testing

Testing is not just executing the documented test cases. There are a multitude of tasks involved in testing, as follows:

- Discussion with product and business about the acceptance criteria

- Creation of a test plan and strategy

- Review of the test plan with the engineering and product teams

- Cross-team collaboration for the test environment setup

- Creation of test cases and test data

- Execution of test cases

- Reporting of test results and associated quality metrics

Depending on the team size, capacity, and structure, some or all of these activities must be performed for a successful release of a software product.

In the following diagram, we get a comprehensive view of all the deliverables involved in software testing:

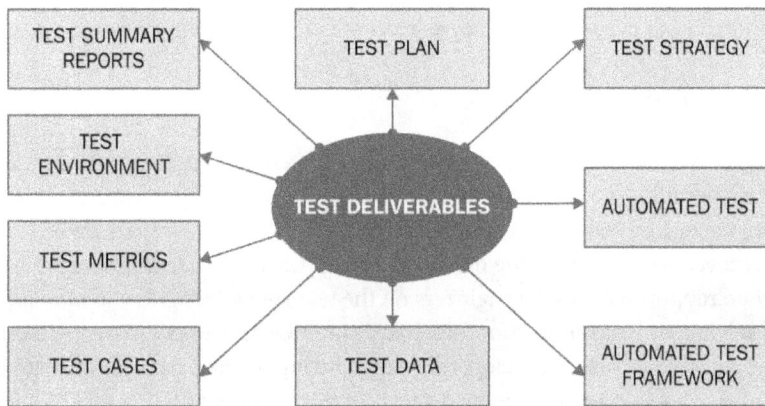

Figure 1.1 – Testing deliverables

As you can see in the preceding diagram, testing encompasses a wide variety of deliverables resulting from many cross-functional activities. Next, let's look at some unique demands for testing in the world.

Testing in an world

Unlike a traditional **waterfall model**, in the Agile world, it is recommended that each user story or feature has the right balance of manual and automated tests before calling it *done*. This arguably slows down the team, but it also drastically reduces the technical debt and improves code quality. You will notice that scaling the software becomes easy, and there is a significant decrease in reworking as the automated tests increase. This saves huge amounts of time for the engineering team in the long run. Another common occurrence in an Agile setup is developers writing automated tests. Usually, developers own unit tests, and quality engineers write other types of tests. Sometimes, it is productive for the developers to write automated test code while the quality engineers focus on other testing tasks. One of the most important takeaways from the Agile world is that quality is everyone's responsibility on the team. Test engineers keep the communication and feedback going constantly between the product and the software engineers, which, in turn, results in an excellent team culture. This also opens up early discussions on the necessary tools and infrastructure needed for testing.

Since the Agile environment aims to deliver the simplest possible product to the customer as quickly as possible, the test engineers focus on the most necessary set of test cases for that product to be functionally correct. In each increment, the test engineers create and execute a basic set of test cases for the features being delivered to the customer. Test engineers are constantly working with the product, developers, and in some cases, with customers to keep the stories as simple and concise as feasible. Additionally, the Agile landscape opens doors to automating the entire software development life cycle using the principles of continuous integration, which demands a major shift in the test engineer's mindset. This demands excellent communication skills and fluent collaboration from the test engineers.

Often, a test engineer's output is measured by the number and quality of defects they find in the software application. Let us examine some key aspects of defect management next.

Defect management in testing

A defect is primarily a deviation from the business requirement, and it does not necessarily mean there is an error in the code. The analyzing, documenting, and reporting of defects in the product is a core activity for the testing team. It is essential to set and follow standard templates for reporting defects. There are a variety of tools in the market that help with test case and defect management. Maintaining a good rapport with all the engineers on the team goes a long way in making the defect reporting and resolving process much smoother. Usually, testers have to put in as much information as possible in the defect log to assist developers in reproducing the defects in various environments as needed. This also helps in categorizing the defects correctly. There are cases when the defect still cannot be reproduced consistently, and the engineer and tester have to pair up in order to spot the failure in the application precisely. It is important to file high-quality functional defects and follow up when necessary to get them fixed on time. So, the testing team is primarily responsible for the defect management process and keeping the higher management updated on any high-severity defects that might affect the product release.

> **Defect versus bug**
>
> A defect is a deviation from the expected behavior. It can be a missing, incorrect, or extra implementation. In comparison, a bug is a programming error that causes the software to crash, produce incorrect results, or perform badly.

Besides the number of bugs being identified, it is also crucial to know when a defect is identified in the project. Next, let's explore the effects of shifting testing early or late in the development life cycle.

Shift-Right and Shift-Left propositions

It is evident that there are a lot of benefits that can be reaped by shifting the testing and quality efforts, in general, earlier on in the development life cycle. This is termed the Shift-Left approach. In this approach, quality engineers are involved right from the inception of the project in early discussions about the design and architecture. They start working on deliverables such as the test plan and test cases in parallel with developers. This approach builds a quality-first mindset for the entire team and ensures quality is built right into the vein of every activity in the development life cycle.

The Shift-Right approach is an extension of Shift-Left, where the testing team's responsibilities are stretched further into the release of the product and maintenance. Test engineers, with their deep knowledge of the product, assist with the implementation and help with testing and monitoring in production. Additionally, the Shift-Right approach has some test engineers support performance, load, and security testing. There is also good scope for test automation to address quality issues under real-world conditions within this approach. Both these approaches are equally important. While the push for Shift-Left has been happening for quite some time now, the Shift-Right approach is relatively new and is extending the tester's involvement in obtaining and testing actionable production data in safer ways.

This leads us to the question of how to incorporate quality early on, and DevOps processes have helped enormously in that regard. Next, let's look at how quality and DevOps blend together to deliver organizational value.

Quality and DevOps

The previous section about the Shift-Right approach leads us right into the area of DevOps and how it has accelerated product development and changed the quality process. DevOps strives continuously to align high-performing engineering teams to business value in the most efficient way. Quality is a core component in each of the DevOps processes. DevOps attempts to automate each and every task in the product delivery, right from building the code to deploying the application to production for customers to use. This adds further emphasis on quality. Since the whole process is automated, feedback at every checkpoint is crucial. A predetermined set of units, functional, and integration tests executed at the right times in the deployment pipeline act as a gate to the production deployment. At times, the manual involvement of test engineers will be required for debugging test failures and for specific types of testing.

In the DevOps world, it is essential to maintain nimbleness in testing activities and deploy the right type of testing resources when and where needed. We will dive deep into how test automation helps in continuous integration and continuous delivery in *Chapter 9, CI, CD, and Test Automation*. For now, it is good to grasp the significant role that testing plays in ensuring a successful DevOps implementation.

So far, we have seen various aspects of testing, and it is undoubtedly a demanding activity. Let's examine some challenges that test engineers face on a daily basis.

Challenges in testing

Before diving into the world of test automation, it is vital to understand the common challenges faced in the testing world. Building a solid manual testing foundation is paramount to developing a sound test automation strategy. A clear and concise testing process acts as a strong pillar for building reliable automation scripts. Some of the challenges faced regularly by the testing team are outlined as follows:

- The most common challenge faced by Agile teams is the changing requirements. Various reasons such as late feedback from customers or design challenges lead to a change in the requirements and cause major rework for test engineers.

- Test engineers are required to interface constantly with the various teams to get the job done. The lack of proper communication here could lead to major blockers for the project.

- Having stories or requirements with incomplete information is another area that test engineers struggle to cope with. They constantly have to reach out to products for additional information, which causes a lot of back and forth.

- Not having the right technical skills for performing all kinds of testing required for a feature poses a great challenge for test engineers.

- The lack of quality-related metrics hurts fast-growing teams where the velocity is expected to increase with the team size. The engineering team will be imperceptive to any patterns in code issues.

- Inadequate test environments are a major bottleneck to testing efforts and directly affect the confidence in the product delivery. The lack of diverse test data in the test environment leads to an inadequately tested product.

- The absence of standard test processes increases both the development and testing times, thereby delaying the project timelines. It is good to be cautious about partially implemented test processes as they, sometimes, hurt more than they help.

Next, let's look at how testing early and often helps overcome some of the obstacles in the testing journey.

Test early, test often

Usually, an enterprise software application contains hundreds, if not thousands, of components. Software testing has to ensure the reliability, accuracy, and availability of each of these components. So, it is inevitable to emphasize the importance of the common quality industry term, *Test early and test often*:

TEST EARLY	TEST OFTEN
• More time to fix bugs	• Drastically increases chance of catching bugs faster
• Cost of fixing bugs is low early on	• Exposes bugs throughout the development life cycle
• Fewer surprises in the later stages of product development	• Easy to make design changes upfront

Table 1.1 – Importance of testing early and testing often

The following excerpt highlights the cost of fixing bugs later in the product development cycle. As Kent Beck explains in his book, *Extreme Programming Explained*, "*Most defects end up costing more than it would have cost to prevent them. Defects are expensive when they occur, both the direct costs of fixing the defects and the indirect costs because of damaged relationships, lost business, and lost development time.*"

The common arguments for not testing early are "*We don't have the resources*" or "*We don't have enough time.*" Both of these are results of not examining the risks of the project thoroughly. It is a proven fact that the total amount of effort is greater when testing is introduced later on in the project. On the tester's part, it is important to reduce redundancy and constantly think about increasing the efficiency of their tests. **Test-Driven-Design** (**TDD**) is a common approach to practice *Test Early, Test Often*. Testing processes have to be fine-tuned and adhered to strictly for this approach to be successful. Testing can have a strategic impact on the quality of the product if introduced early and done often and efficiently. The Agile test pyramid (which is discussed, in detail, in *Chapter 2*) can act as a guide to strategically categorize and set up different types of tests.

So far, we have dealt with a range of concepts that help us acquire a well-rounded knowledge of testing. We looked at the importance of testing and how crucial it is to integrate it with every facet of the software development process. With this background, let us take on the primary topic of this book, test automation.

Understanding test automation

Imagine an engineering team adding one or two software engineers every quarter. This team wants to delight the customer by delivering more features every sprint. Even though they have one or two quality engineers on the team to test all of the new features, they notice that the faster they try to deliver code, the higher the number of regression bugs introduced. The manual testing of the features just isn't enough. They want certain core test scenarios to be executed repetitively and, often, when the changes are introduced. This is where test automation helps tremendously. It is always better to get started with test automation before feeling the agony of a high-severity bug in production or a catastrophic incident happening due to the lack of timely testing.

Testing is definitely not a one-time activity and must be done any time and every time a change is introduced into the software application. The longer we go without testing an application, the higher the chances of failure. So, continuous testing is not an option but an absolute necessity in today's Agile software engineering landscape. Manual testing involves a tester executing the predefined test cases against the system under test as a real user would. Automated testing involves the same tests to be run by a script, thereby saving the tester's valuable time so that they can focus on usability or exploratory testing. When done right, automated tests are more reliable than a manual tester. More importantly, it provides more time for the tester to draw valuable insights from the results of the automated tests, which further aids in increasing the test coverage of the software application as a whole.

Test automation is one of the chief ways to set up and achieve quality in an orderly fashion. The core benefit of test automation lies in identifying software bugs and fixing them as early and as close as possible to the developer's coding environment. This helps subdue the damaging effects of late defect detection and also keeps the feedback cycle going between the engineering and product teams. Even though the upfront investment in automated tests might seem large, analysis has shown that, over time, it pays for itself. Test automation enables teams to deliver new features quickly as well as with superior quality. *Figure 1.2* shows how various components of a software application can be interdependent, and continuously testing each of them asserts their behavior. It is not only necessary to validate these components as a standalone but also as an integrated system:

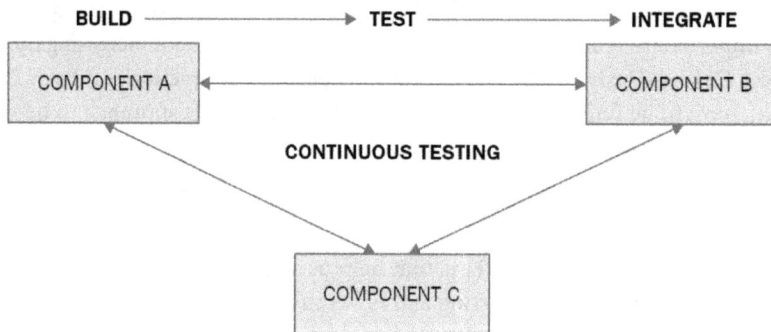

Figure 1.2 – Continuous testing

Imagine a simple loan application flow created by a combination of three **Application Programming Interfaces (APIs)**. The individual APIs, in this case, could be as follows:

1. Creating the applicant

2. Creating a loan application

3. Determining loan eligibility

Each of the APIs has to be tested in an isolated manner at first to validate the accuracy of the business functionality. Later when the APIs are integrated, the business workflow should be tested to confirm the system behavior. Here, a test automation suite can be built to automate both the test cases for individual APIs and the whole system behavior. Also, when the applicant creation API is reused in another application, the automation suite can be reused, thus enabling reusability and portability in testing. A similar implementation can be done for user interface components, too.

Test automation is a very collaborative activity involving the commitment of business analysts, software engineers, and quality engineers/**software development engineer in test (SDET)**. It unburdens the whole team from the overwhelming effects of repetitive manual tests, thus enabling the achievement of quality at speed. While peer code reviews and internal design reviews act as supplemental activities to identify defects early, test automation places the team in a fairly good place to start testing the product with the end users. It is a common misconception that automated tests undercut human interaction with the system under test. While it might be true that the tester does not interact with the system as often as they would in manual testing, the very activity of developing and maintaining the automated tests brings together the whole team by commenting on the test code and design. Automated tests open a new way of communication within the team to improve the quality of the system and the prevention of bugs.

Having looked at the basics of test automation, let's learn some important considerations in the Agile world.

Agile test automation

Having the right selection of automated tests at the right spots in the deployment makes a ton of difference in the quality of the delivered software. In an Agile environment, there has to be a constant discovery of the right tests for the current iteration, and the ratio between the manual and automated tests needs to be tweaked as and when necessary. Since the focus is on delivering a feature as quickly as possible to the customer, it is important that developers, testers, and the product manager are aware of what is being built, tested, and shipped. Collaboration becomes crucial and is the primary driver of the success of test automation in the Agile environment. Some of the important considerations for test automation in an Agile environment are as follows:

- Start small and build iteratively on the automated test scripts. This applies to both functional test coverage and the complexity of the test automation framework.

- Be extremely cautious about what tests are selected for automation. These tests will act as a gate to production deployment. Make every effort to avoid false positives and false negatives.

- Make sure the automated tests are considered when deciding on the acceptance criteria for a feature or a story. It is critical to allocate necessary time and resources for completing and executing the automated tests as part of a feature.

- Get frequent feedback from other engineers on the quality and performance of the automated tests.

- Do not be afraid to adapt and change test automation. Constant innovation and improvement are core activities in an Agile environment.

As we have seen so far, test automation is an intricate pursuit. Like all other complex things in the software world, test automation comes with its own set of challenges. Let's examine some of them in the next section.

Test automation challenges

There is never enough test automation, and this is a constant challenge faced by test engineers. Test engineers are always under time constraints to finish the manual or automated tests and to get the completed feature out of the door. Just as with manual testing, there are a variety of challenges that test engineers face on a daily basis. Some of them are listed as follows:

- As mentioned earlier, not having enough test automation is always a challenge in the Agile world. Often, testers have to compromise coverage for speed, and at times, the consequence can be adverse.

- Not enough planning before beginning test automation work often leads to duplication and exhaustion.

- Upfront investment in test automation is heavy, and the constant need to convince the stakeholders of the benefits puts a lot of pressure on testing teams.

- The lack of collaboration between developers and test engineers while designing and developing automated tests could lead to poor-quality scripts or complicated frameworks. This affects the quality of the build pipeline.

- The absence of skilled test engineers has a detrimental effect on quality.

- Not aligning the testing processes with the development processes could cause release delays. It is extremely important to time the code completion of the feature with automated test readiness for on-time delivery.

- Sometimes, test engineers do not understand the requirements and assume the expected behavior instead of validating it with the product. It is extremely hard to identify and eliminate such assertions in the test script after they have been built in.

- The test automation framework is not scalable or portable. This hinders the execution speed and results in a lack of test coverage.

Creating and maintaining the test automation infrastructure is a separate project by itself. Every bit of detail should be thought about and discussed in detail with the concerned stakeholders to address these challenges.

Finding and handling regression bugs

One of the chief purposes of test automation is finding regression bugs. Usually, regression means the quality gets worse after a change has been introduced to an already tested product. Well-written automated tests are tremendously helpful in identifying regression bugs. But an important thing to remember about regression issues is that even though more automated tests help find them, the root cause of regression issues still has to be addressed by the management team. Test engineers can provide awareness about the inherent regression issues through automated test results. It is important to educate the team about fixing the underlying lapses. Some of the most common slippages that cause regression bugs include the following:

- Code review standards are below par or non-existent.
- The project schedule is no longer realistic to push good quality code through.
- There is a huge disconnect between developers and the product teams regarding the feature being developed.
- Various integration points are not being considered appropriately when designing the product.
- The product has become too complex over time.

Test engineers can be a key component to break the pattern of regression issues and steer the quality of the product in the right direction by keeping everyone on the team informed. In the next section, let's survey some of the top metrics used in the test automation world.

Test automation metrics

Why and what to measure in test automation is a constant question lingering in the minds of the testing team. Test engineers are curious to know how effective their scripts are and how they are performing across different conditions. On the other hand, the management team will be interested to know the ROI on the investment made in test automation rather than just hoping that they deliver value in the long run. Let's look at some key metrics that can help in gauging the value of test automation that is already in place:

- **Test automation effectiveness**: This is a key metric that provides visibility into how effective the automated tests are in finding bugs. When broken down logically by scripts/test environments, this metric gives direction on where to focus our future efforts:

```
Test Automation Effectiveness = (# of defects found by automation/
Total number of defects found) * 100
```

- **Test automation coverage**: This metric provides the test coverage for the automated tests. It is important to be aware of what is not covered by your test automation suite to clearly direct the manual testing efforts:

  ```
  Test Automation Coverage = (# of test cases automated/Total
  number of test cases) * 100
  ```

- **Test automation stability**: This metric provides clarity on how well the automated scripts are performing in the long run. This metric, along with test automation effectiveness, is a key indicator of how flaky the automated tests are. It is good to have this metric in an Agile environment to monitor the health of the deployment pipelines:

  ```
  Test Automation Stability = (# of test failures/Total number
  of test runs) * 100
  ```

- **Test automation success or failure rate**: This metric provides a quick idea of the health of the current build. In the long term, this metric gives you a good view of how the build qualities have changed:

  ```
  Test Automation Success Rate = (# of test cases passed/Total
  number of test cases run) * 100
  ```

- **Equivalent manual test effort**: This metric, usually expressed in man hours, is an indicator of the ROI to determine whether a particular feature can be automated or not:

  ```
  Equivalent Manual test Effort = Effort to test a specific feature
  * Number of times a feature is tested in a test cycle
  ```

- **Test automation execution time**: As the name indicates, this metric shows you how long or short your test runs are. Comparing this with the overall deployment times to production gives a good idea of how much time is spent executing the automated tests over time. This is key in the Agile setup where the speed to the customer is considered a key factor:

  ```
  Test Automation Execution Time = Test Automation End Time - Test
  Automation Start Time
  ```

An important note regarding test automation metrics is that these numbers, by themselves, do not produce much value. A combination of these metrics properly assessed in the context of project delivery provides valuable insights to the product and engineering teams.

So far, we have seen how testing and test automation blend well with every activity in the software development life cycle. It is evident that test automation is a multidimensional undertaking, and the people performing test automation have to meet the needs of this demanding role. In the next section, let's survey a couple of important roles in the quality engineering space.

Exploring the roles in quality engineering

There are three main roles in the testing world without involving people management. Different companies might use or call these roles in different ways. Quality-related roles also vary depending on the size and structure of the organization. For example, smaller companies might combine all of these into a single role and call it SDET. The main differences stem from their technical expertise and overall experience in the software and quality engineering domains:

- Traditional manual tester (or quality assurance analyst)

- Test automation engineer (or software quality engineer)

- **Software Development Engineer in Test (SDET)**

The traditional manual tester role is not as prevalent as it used to be, as every role in quality is expected to perform some level of test automation in the Agile world. So, here, we will be focusing only on the latter two. This does not mean that manual testing is not done anymore. The responsibilities of a manual tester are shared by test engineers, SDETs, business analysts, and product owners/managers.

There are organizations where software engineers/developers perform their own testing. The engineer developing the feature is responsible for performing the required testing and delivering a high-quality product. Often, small- to mid-sized companies use this approach when they are still not able to build a separate quality organization or hire test engineers. An important downside of this is that there is a lot of context-switching that needs to be done by the developers. Also, consistently maintaining the test coverage at high levels becomes a challenge. A single developer will be focused on providing good coverage for their own feature, and it is very easy to miss the integration or system-level tests needed for a business workflow. Some developers, plain and simple, do not like to perform tests other than unit tests due to the tedious and repetitive nature of testing. It is also hard to keep track of the various details involved in testing different parts of an application.

Next, we will look at the most common testing roles that are prevalent in the market.

Test automation engineer

Test automation engineers are the core members of the quality organization within the software engineering teams. They can either be embedded into the engineering team or be part of a team of quality engineers reporting to a quality manager. The main responsibilities of a test automation engineer include the following:

- Test planning and test strategy development for testing product features (both manual and automated)

- Preparation of test cases and test data

- Setting up the test environment (usually with help from other engineers)

- Create, execute, and maintain automated tests for the product features

- Uses the existing test automation infrastructure to build a sound test automation strategy

- Collaborate with product and implementation teams to achieve good test coverage

- Reporting and retesting of bugs

- Coordinating bug fixes across teams if necessary

- Streamline test processes

To sum it up, on a daily basis, test engineers have to partner with most of the members of the team and stay on top of the user stories. They take ownership of the product quality of the team they are embedded into. Next, let's see how an SDET is different from a test automation engineer.

SDET

An ideal candidate for the SDET position exhibits sound technical skills and a deep expertise in testing methodologies. For all practical purposes, technically, an SDET is as good as a software engineer with extensive knowledge of the quality engineering space. An SDET will be involved throughout the development life cycle from unit test creation to production release validation and always strives to enhance the productivity of both the software engineer and quality engineer in the team.

The main responsibilities of an SDET include the following:

- Setting clear objectives for test automation

- Creating and improving the test automation infrastructure

- Owning the test automation strategy

- Liaison with software engineers on the team and across teams (if needed) to build and maintain the automation framework

- Being involved in the design and architectural discussions

- Acting as a mentor for the quality engineers in the team

- Interfacing with the DevOps team to ensure testing happens at every stage of the development pipeline

- Adapting and implementing the latest technological developments in the quality engineering domain

Table 1.2 highlights the differences between a test automation engineer and an SDET:

Test automation engineer	SDET
Creates and executes automated and manual tests	Creates and maintains the test automation framework
Collaborates with the product and implementation teams	Collaborates with software engineers and DevOps teams
Highly skilled in programming with testing skills	Experts in testing either manually or by automation
Develops test automation tools	Uses test automation tools

Table 1.2 – Test automation versus SDET

Even though their roles demand different responsibilities, both the quality engineers and SDET are equally accountable for the release of a bug-free product to the customer. Both should be in the meetings with product stakeholders to make a final decision on every feature release. A quality engineer is good at test case creation, while an SDET specializes in choosing the right ones to automate in the best possible way. There are times when quality engineers and SDETs have to work in tandem to keep the upper management informed and educated about the capabilities of test automation and the effort it takes to achieve the ROI. Also, it is important to note that the relationship between software engineers and quality engineers/SDETs is of utmost importance to the success of any test automation work. It is vital to get continuous feedback from software engineers on the test automation code and design. Software engineers should also be educated, when necessary, about the various benefits derived from test automation

In the next section, let's get ourselves familiarized with some commonly used definitions in the world of testing and test automation.

Familiarizing yourself with common terminologies and definitions

In this section, we will look at some of the most commonly used terms in the quality engineering space. Since quality engineering is part of the software engineering practice, you would notice quite a few familiar terms if you're a software engineer:

- **A/B testing**: Testing to compare two versions of a web page for performance and/or usability to decide which one better engages the end users.

- **Acceptance testing**: A testing technique to establish whether a feature/product meets the acceptance criteria of business users or end users.

- **Agile methodology**: This is an iterative approach to software development that puts collaboration and communication at the forefront. Essentially, it is a set of ideas to deliver software to customers quickly and efficiently.

- **Behavior-Driven Development (BDD)**: BDD is a common Agile practice where critical business scenarios are first documented and then implemented to make sure the end product evolves continuously with shared understanding. We will look at an implementation of the BDD framework in *Chapter 7*.

- **Black-box testing**: This testing technique focuses on the output of a product, ignoring the design and implementation details.

- **Data-driven testing**: A common automated testing approach where a reusable logic, often part of a test script, is run over a collection of test data to compare the actual and expected results.

- **End-to-end testing**: A testing technique to ensure that the integrated components of a product work as expected. This is a very important testing type to verify critical business flows.

- **Exploratory testing**: A manual testing approach where the product is tested in an investigative and inquisitive manner without documented steps with the main goal of finding bugs.

- **Integration testing**: A testing technique in which the communication logic of the individual software components or services are combined and tested as a group.

- **Load testing**: A type of testing to evaluate the performance of a software application by simulating the real-world traffic on the system. It can be done with a software system as a whole or just an API or database. We will consider the setup for load testing in *Chapter 8*.

- **Penetration testing**: This is a type of security testing where a tester attempts to find and exploit the vulnerabilities of a software application.

- **Security testing**: A testing type to ensure that all the required defenses are in place against various types of cyber threats.

- **Stress testing**: This is a type of performance testing where the system is subject to heavy loads with the intention of breaking it. Stress testing is mainly used to determine the performance limits of a software application. We will look into the setup for stress testing in *Chapter 8*.

- **API testing**: This is a testing type focused on verifying the API's logic, build, and structure. The main goal is to validate the functional logic of the APIs. We will look at an implementation of API testing in *Chapter 7*.

- **Smoke testing**: A testing technique that is primarily used to check the core features of a product when there is a change introduced or before releasing it to a wider audience.

- **System testing**: A testing type used to evaluate the software system as a whole to make sure the integration of all the components is working as expected.

- **Test case**: This is a collection of steps, data, and expected results to test a piece of code or a functional component within a software application.

- **Test plan**: This is a document that outlines a methodical approach to testing a software application. It provides a detailed perspective on testing the various parts of the application in its parts and as a whole.

- **Test-Driven Development (TDD)**: This is an approach mainly used in the Agile world where a test is written first followed by just enough code to fulfill the test. The refactoring of the code continues to follow the tests, thus keeping the emphasis on specifications.

- **Test suite/test automation suite**: A collection of automated test scripts and associated dependencies used to run against the software application.

- **Unit testing**: A type of testing in which the logic of the smallest components or objects of software is tested. This is predominantly the first type of testing performed in the development life cycle.

- **Usability testing**: This is a testing technique used to evaluate a product (usually, a web application) with the aim of assessing the difficulties in using the product.

- **Validation**: This is the process of ensuring the input and output parameters of a product. It answers this question: *Are we building the product correctly*?

- **Verification**: This is the process of checking whether the product fulfills the specified requirements. Verifications answer question: *Are we building the right product?*

- **Waterfall model**: This is a project management methodology in which the software development life cycle is divided into multiple phases, and each phase is executed in a sequential order.

- **White-box testing**: A testing technique to validate the features of a product through low-level design and implementation.

We have equipped ourselves with the knowledge of a wide array of terms used in the quality industry. Now, let's quickly summarize what we have seen in this chapter.

Summary

In this chapter, we looked at the importance of testing and test automation. We dealt with some practical guidance on how test automation interfaces with different teams and all the members of the engineering team. Additionally, we looked at the roles in the test automation world and their specifics and similarities. We understood how test automation helps achieve quality at every stage of the product development life cycle. Finally, we looked at the different terminologies used in test automation.

In the next chapter, we will examine what a thorough test automation strategy entails and how to define one. Additionally, we will review a few common test automation design patterns.

Questions

Take a look at the following questions to test your knowledge of the concepts learned in this chapter:

1. What is software testing, and why do you have to do it?
2. What are some important deliverables in software testing?
3. What is test automation, and how does it help testing?
4. What are the challenges faced in testing and test automation?
5. What are the common roles in quality engineering?

Test Automation Strategy

Effective test automation requires creativity, technical acumen, and persistence. Success is achieved in test automation by persevering through a wide variety of challenges posed by complex application areas, changing requirements, and unstable environments. Every successful test automation project starts with a sound strategy. Every software project is unique, and a test automation strategy serves as a beacon to meet the unique requirements of a test automation project. It acts as a pillar to inquire, assess, and make constant improvements to the overall software quality.

Test automation strategy is not only about planning and executing the automated tests but also ensuring that by and large the efforts in test automation are placed in the right areas to deliver business value quickly and efficiently.

We will be looking at the following topics in this chapter to solidify our understanding of test automation strategy and frameworks:

- Knowing your test automation strategy
- Devising a good test automation strategy
- Understanding the test pyramid
- Familiarizing ourselves with common test automation design patterns

Technical requirements

In the later part of this chapter, we will be looking at some Python code to understand a simple implementation of design patterns. You can refer to the following GitHub URL for the code in the chapter: `https://github.com/PacktPublishing/B19046_Test-Automation-Engineering-Handbook/blob/main/src/test/design_patterns_factory.py`.

This Python code is provided mainly for understanding the design patterns and the readers don't have to execute the code. But if you are interested, here is the necessary information to get it working on your machine. First, readers will need an **integrated development environment** (**IDE**) to work through the code. **Visual Studio Code** (**VS Code**) is an excellent editor with wide support for a variety of programming languages.

The following URL provides a good overview for using Python with VS Code:

`https://code.visualstudio.com/docs/languages/python`

You will need software versions Python 3.5+ and the **Java Runtime Environment** (**JRE**) 1.8+ installed on your machine to be able to execute this code. `pip` is the package installer for Python, and I would recommend installing it using `https://pip.pypa.io/en/stable/installation/`. Once you have PIP installed, you can use the `pip install -U selenium` command to install Selenium Python bindings.

Next is to have the driver installed for your browsers. You can do this by going to the following links for your particular browser:

- Chrome: `https://chromedriver.chromium.org/downloads`
- Firefox: `https://github.com/mozilla/geckodriver/releases`
- Edge: `https://developer.microsoft.com/en-us/microsoft-edge/tools/webdriver/`

Make sure the driver executables are in your `PATH` environment variable.

Knowing your test automation strategy

Even though every software project has its own traits, there are certain important aspects that every test automation strategy should touch upon. Let us look at some chief considerations of test automation strategy in this section.

Test automation objectives

One of the first tasks that an automation strategy must define is its purpose. All the project stakeholders should be involved in a discussion about which pain points test automation would address eventually. This step is also crucial because it assists management to redirect resources to the appropriate areas. Some of the most common objectives for test automation are noted here:

- **Strengthening product quality**: This is a chief trigger for test automation projects. Test automation engineers can plan and create tests that can catch regression bugs and can help with defect prevention.
- **Improving test coverage**: This is a key motivation for several test automation undertakings. Test engineers usually work with the software engineers and the product team to cover all possible scenarios. The scenarios can then be categorized based on their criticality.

- **Reducing manual testing**: Needless to say, reducing manual testing is a primary objective for most—if not all—test automation initiatives. Automated tests are predictable and perform the programmed tasks in a consistent and error-free manner. This leaves more time for the whole team to focus on other important tasks, and it boosts the overall productivity of the team.

- **Minimizing maintenance and increasing portability of tests**: This can be an incredibly good reason for kicking off a test automation initiative. Legacy software projects can be ripe with tightly coupled tests that are not compatible with newer technologies. It usually requires considerable effort from the testing team to make them portable and easy to read.

- **Enhancing the stability of a software product**: This acts as a principal motivation to kick start a test automation project to address non-functional shortcomings. This involves identifying and documenting various real-life stress- and load-testing scenarios for the application under test. Subsequently, with the help of the right tools, the application can be subject to performance testing.

- **Reducing quality costs**: This can help the management invest their resources and funds to further the capabilities of the organization. Reducing redundancy in quality-related efforts helps trim testing costs and time. Test automation strategy can be a great driver for this endeavor.

A push for test automation can be due to one or a combination of these reasons. Well-thought-out objectives for test automation eventually increase the overall efficiency of the engineering team, thereby aligning their efforts toward achieving greater business value.

Next, let us look at how important it is to garner management support to achieve the objectives of test automation.

Gathering management support

One of the critical factors that influence the outcome of a test automation effort is constant communication with management about the progress and issues. Isolated efforts from teams in building test automation might work in the short term, but to scale up the infrastructure needed to serve the entire organization, we need adequate support from management. Calling attention to the business goals that test automation can help achieve is one of the ways to engage management. It is crucial to relay to management how test automation unites the divide between frequent release cycles and a bug-free product. Having a clear test automation strategy helps exactly with that. Management will have clarity on what to expect in the initial phase when the framework is being built and when they can start seeing the **return on investment (ROI)**. Another factor to be highlighted is that test automation, once put in place, increases the efficiency of multiple teams.

After laying out the objectives and getting adequate support for the test automation effort, let us next deal with defining the scope as well as addressing some limitations of test automation.

Defining the scope of test automation

The scope of test automation refers to the extent to which the software application will be validated using automated tests. There are a number of considerations that go into determining the scope. Some important ones are listed here:

- Number of business-critical flows

- The complexity of test cases

- Types of testing to be performed on the application

- The skill set of test automation engineers

- Reusability of components within the application

- Time constraints for product delivery

- Test environment availability/maturity

It is unrealistic to plan to automate every part of a software application. Every product is unique, and there is no absolute path to determine the scope of test automation. Planning upfront and constant collaboration between test engineers, software engineers, and product owners help lay out the scope initially. Starting small and building iteratively assists in this approach. It is also crucial to distinguish upfront which types of testing will be done and by who.

Let us now investigate the crucial role that the test environment plays in a test automation strategy.

Test automation environment

Where and how automated tests are executed forms an integral part of a test automation strategy. A test automation strategy should address deliberate actions for correctly setting up a test environment and monitoring for active use. There should also be an effective communication strategy to provide timely updates on test environments. Now, let us look at what really constitutes a test environment.

What constitutes a test environment?

The infrastructure team usually makes a replica of the code base and deploys it on different **virtual machines** (**VMs**) with the associated dependencies. They provide access to these VMs for the required users. Users can then log on to these and perform a series of operations based on their roles and permissions. The combination of the hardware, software, and state of these code bases makes it a separate test environment.

The most common types of test environments are presented here:

- Development
- Testing
- Staging
- Production

We will not be going into the details of each type here, but it is important to keep in mind that the testing environment is where most of the automated tests are run. The staging environment is usually used for **end-to-end** (**E2E**) testing and **user acceptance testing** (**UAT**). This can be both manual and automated. All organizations have a development environment, but depending on the size and capacity, some may combine the testing and staging environments into one. Now that we know what constitutes a test environment, let us look at how to provision one.

Provisioning a test environment

A key factor that determines the stability of automated tests is the test environment. Serious considerations must be made in the setup and maintenance of test environments. This is where the test engineers must collaborate effectively with the infrastructure teams and identify the limitations of their tests in the context of a test environment. This analysis, if done early in the project, goes a long way in building resilient automated tests. After all, no one likes to spend a huge amount of time debugging test failures that are not functional in nature.

Test environment provisioning and configuration should be automated at all costs. Manual setup increases the cycle time and adds unnecessary overhead for testers and infrastructure teams. A wide variety of tools such as **Terraform** and **Chef** are available on the market for this purpose. Although responsibility for the provisioning of various environments lies with the infrastructure team, it does not hurt for the test engineers to gain a basic understanding of these tools. An ideal environment provisioning approach would address setting up the infrastructure (where the application will run), configurations (behavior of the application in the underlying infrastructure), and dependencies (external modules needed for the functioning of the application) in an automated manner.

Testing in a production environment

Apart from a few high-level checks, any additional testing in production is usually frowned upon. New code changes are usually soaked in a pre-production environment and later released to production when the testing is complete (manual and automated). But in recent times, *feature flags* are being routinely used to turn off/on new code changes without breaking existing functionality in production. Testing in production can be localized to a minor subset of features by using feature flags effectively. Even if there are features broken by the introduction of new code changes, the impact would be minimal. Test automation strategy helps maintain sanity by providing adequate test coverage for critical features before they are deployed to production. Irrespective of the state of feature flags, primary business flows should be functional, and it is a prime responsibility of test automation to keep it that way.

> **Note**
> Automated tests are only as good as the environment they run against. An unstable test environment adversely impacts the accuracy, efficiency, and effectiveness of the test scripts.

Let us next see how a test automation strategy enables quality in an Agile landscape.

Implementing an Agile test automation strategy

As a majority of organizations have adopted the Agile paradigm, it is mainstream now, and test automation strategy must necessarily address the Agile aspects of software development. Fundamental Agile principles must be engrained in the team's activities, and they should be applied to test automation strategy when and where possible. This might be a huge cultural change for a lot of testing teams. Here is a list of Agile principles for consideration in test automation strategy:

- Gain customer satisfaction through steady delivery of useful software
- Embrace changing requirements to provide a competitive advantage to end users
- Deliver working software frequently
- Constant collaboration between business and engineering teams
- Motivate individuals and trust them to get the job done
- Prefer face-to-face conversation over documentation
- Working software as a primary measure of progress
- Promote a continuous and sustainable pace of development indefinitely
- Continuous attention to technical excellence and superior design
- Keep things simple
- Each team is self-organized in its design, architecture, and requirements
- Continuous retrospection and improvements

A primary goal in the Agile journey is to enhance the software quality at every iteration. Test automation strategy should have automated builds and deployment as a pre-requisite. Only when build automation is coupled with automated tests is continuous improvement enabled. This is how teams can maximize their velocity in the Agile landscape while maintaining high quality.

Another key aspect to zero in on is the ability to get feedback early and often. It helps tremendously to instill a test-first mentality in every member of the team. When an engineer thinks through the feature and which kinds of tests should pass before writing any code, the output is bound to be of high quality. This also keeps the team from accruing any technical debt in terms of test automation.

Attentive and incremental investments in test automation should be the foundation of the strategy, and it goes a long way in tackling the frequent code changes in Agile development. Sprint retrospections can be used to pause and reflect on areas where Agile principles are not being effectively applied to test automation strategy.

Agile testing quadrants provide a logical way to organize different kinds of tests based on whether they are business-facing or technology-facing. Let us quickly look at which types of tests are written in each of the quadrants, as follows:

- **Quadrant 1**: Highly technical tests written by the programmers, usually in the same programming language as the software application.

- **Quadrant 2**: Tests that verify business functions and use cases. These tests can be automated or manual and are used to confirm product behavior with the specifications.

- **Quadrant 3**: Manual tests that verify the **user experience** (**UX**) and functionality of the product through the eyes of a real user.

- **Quadrant 4**: Tool-intensive tests that verify non-functional aspects of the application such as performance and security.

Test automation strategy should also address the diverse needs for tools and processes in each of the Agile testing quadrants, outlined in the following diagram:

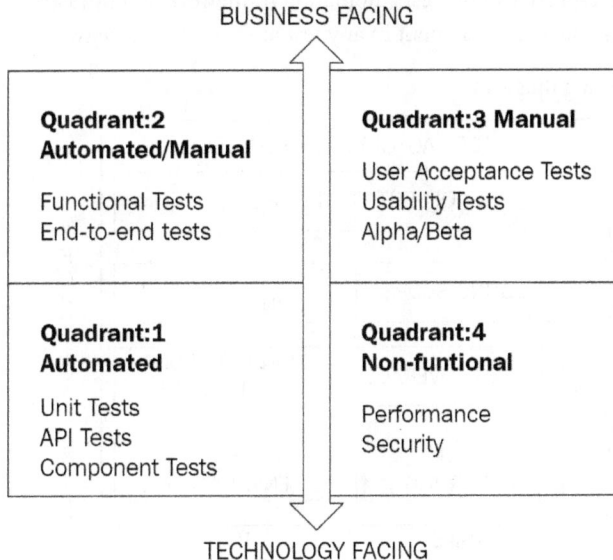

BUSINESS FACING

Quadrant:2 Automated/Manual Functional Tests End-to-end tests	Quadrant:3 Manual User Acceptance Tests Usability Tests Alpha/Beta
Quadrant:1 Automated Unit Tests API Tests Component Tests	Quadrant:4 Non-funtional Performance Security

TECHNOLOGY FACING

Figure 2.1 – Agile testing quadrants

Test automation strategy in an Agile project should promote team collaboration, thus acting as a strengthening force to achieve superior product quality. Next, let us look at some reporting guidelines to be incorporated into the test automation strategy.

Reporting the test results

There are usually thousands of tests that run part of a pipeline in an enterprise setup. This is just a small number when compared to some advanced tech-savvy organizations that have millions of individual tests run daily. Even a small percentage of failure translates to a massive debugging and maintenance effort. Test automation strategy should incorporate effective test grouping in case of failures. There could be several reasons for a test failure, including—but not limited to—new bugs, flakiness of the test script, test environment issues, and so on. Analyzing failures and producing a quick fix wherever necessary is principal to maintaining a working product.

Test automation strategy should enable the labeling of test failures and open appropriate channels to communicate the failures. Integration with enterprise message software helps in reporting quickly and effectively. This is also an area ripe for **machine learning** (**ML**) usage in the quality engineering space.

There should also be a single portal where the test results can be filtered and consumed for providing visibility to interested stakeholders.

Having the right reporting in place for test automation provides a major boost to the productivity of the engineering team. Irrespective of the test automation framework and infrastructure, standardizing test results reporting is a must-have element in any test automation strategy.

Have a look at the following diagram:

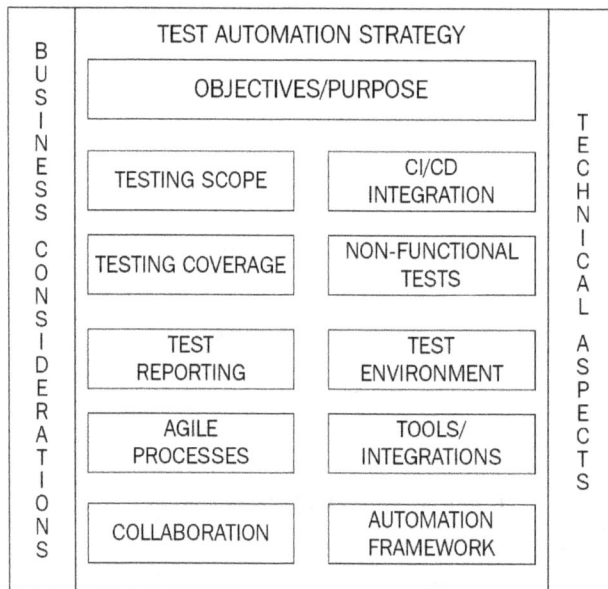

Figure 2.2 – Test automation strategy breakdown

Figure 2.2 provides a view of all the major constituents of a test automation strategy. A good test automation strategy would be a combination of these items based on the individual software application.

Next, let us look at some chief items to consider when devising a good automation strategy.

Devising a good test automation strategy

After defining the objectives for your test automation effort, it is vital to put all the pieces together and hit the road running as soon as possible. The longer you wait to kick start your test automation efforts, the higher the technical debt you accrue. In this section, let us dive deep into some practical tips for implementing and maintaining the key parts of the test strategy.

Selecting the right tools and training

A search for a valuable tool always commences by gaining a deep understanding of the project requirements. The type of application and distinct kinds of platforms to be tested are some chief points of interest to get started. It might help to do a cost-benefit analysis on developing an in-house tool from scratch versus buying a commercially licensed tool. Developing a tool from scratch takes a significant amount of time and requires putting together solutions from multiple open source tools with in-house code modules. This hybrid approach is more cost-effective than a commercial license tool but requires dedicated support from the internal teams to keep up with changes to the external libraries.

For projects that do not have a lot of time upfront, buying the license from a vendor might be the only option on the table. Every test automation tool on the market now has some flavor of cloud offering that can be used to get feet wet initially with a subset of users. Once the pilot is successful, the tool can be rolled out to a broader set of users. There might be some additions and extensions needed to meet specific project needs, even after buying a vendor tool from the market. Look out for additional integrations that the tool will have to support for tracking stories and defects.

Irrespective of the type of tool used, the key factor for success is the training that goes along with it. It is important to educate the test engineers and other users of the tool regarding its effectiveness and limitations. Every user should be equipped with the knowledge and skill set needed to use the tool efficiently. Always try to use a programming language that is most prominent among engineers and testers in the organization. Languages such as Python, Ruby, and JavaScript can be good starting points for novice testers with minimal programming experience. They have tremendous community support and span a wide variety of domains, including the **Internet of Things (IoT)** and ML. It is also important to consider how well the programing language integrates with the numerous tools in the testing ecosystem. The majority of the time, a quality engineer is focused on quickly solving the problem at hand. Selecting the right programming language goes a long way in assisting this purpose. In some cases, there may be non-technical users who wish to use the test automation tool for running acceptance tests. It is critical to keep in mind the ease of usability in such cases.

A test automation tool is only a small part of the test automation strategy. High-priced tools do not guarantee success if other standards of the test automation framework are missing or below par. Next, let us consider the subject of a test automation framework and pay attention to the standards surrounding it.

Standards of the test automation framework

A test automation framework is a collection of tools and/or software components that enable the creation, execution, and reporting of tests. There should also be some guidelines and standards associated with the framework so that every engineer in the organization utilizes and builds on it the same way. It facilitates a systematic way for test engineers to efficiently design test scripts, automate their execution, and report the test results. Test automation is also crucial to improving the speed and accuracy of test execution, thereby ensuring a high ROI on the underlying test automation strategy.

Here are some important test automation framework types to be aware of:

- **Modular test automation framework**: In this framework, the application logic and the utilities that help test the logic are broken down into modules of code. These modules can be built on incrementally to accommodate further code changes. An example would be an E2E test framework where the major application modules have a corresponding test module. Changes to the validation logic happen only within the revised modules limiting rework and increasing test efficiency.

- **Library-based test automation framework**: In a library-based framework, the application logic is written in separate methods, and these methods are grouped into libraries. It is an extremely reusable and portable test automation framework since libraries can be added as dependencies externally. Most test automation frameworks are library-based, and this relieves the engineers from writing redundant code.

- **Keyword-driven test automation framework**: The application logic being tested is extracted into a collection of keywords and stored in the framework. These keywords form the foundation of test scripts and are used to create and invoke tests for every execution. An example of this would be the **Unified Functional Testing** (UFT) tool, formerly known as **Quick Test Professional** (QTP). It has inherent support for keyword extraction and organizing libraries for each of those keywords.

- **Data-driven test automation framework**: As the name suggests, data drives the tests in this framework. Reusable components of code are executed over a collection of test data, covering major parts of the application. A simple example of a data-driven framework would be reading a collection of data from a database or a **comma-separated values** (CSV) file and verifying the application logic on every iteration of the data. There is usually a need to integrate with a plugin or driver software to read the data from the source.

- **Behavior-driven development (BDD) test framework:** This is a business-focused Agile test framework where the tests are written in a business-friendly language such as Gherkin. Cucumber is one of the most popular BDD frameworks on the market.

Have a look at the following diagram:

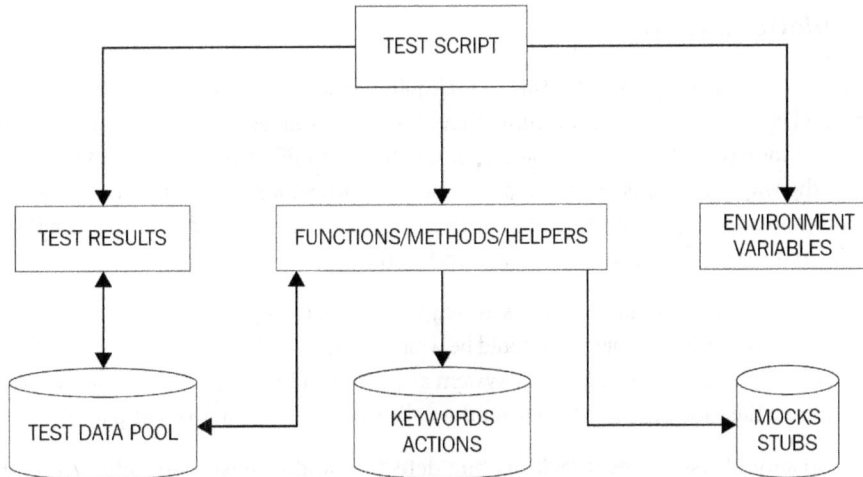

Figure 2.3 – Essential components of a test automation framework

Often, a real-world test automation framework is a combination of two or more of the types shown in *Figure 2.3*. The test script acts as a primary driver to invoke helper methods to perform a variety of tasks. The test script also leverages the environment variables and is responsible for managing the test results.

Now that we know what a test automation framework is, let us next look at some key components of every test automation framework.

Managing test libraries

The principal component of an automation framework is its collection of libraries. An automation framework should empower the test engineers to extract the core application logic and store this as *test libraries*. For example, in a banking application, a loan creation method can be used for all types of testing (unit, integration, E2E). The test automation framework should create this method or inherit it from the source code and make it available for consumption.

The test automation framework should also support the usage of third-party libraries for performing various tasks within the tests. Any third-party library used as a dependency within a framework should carefully be vetted for security and performance concerns. These libraries are often used as utilities that aid test scripts to perform certain recurring tasks.

Both the custom and third-party libraries together form the building blocks of a strong test automation framework. Whichever form they may be in an automation framework, they hold the essential application logic and utilities that assist in validating the logic.

Next, let us look at how important it is to have a good platform for running tests.

Laying the platform for tests

A test automation framework primarily aims to maintain consistency across test scripts. When test scripts are created by a team in the organization, they should be available for reuse for any team in the future. The test automation framework should promote this kind of mobility, which in turn enables consistency in the way test scripts are created. Test automation should create foundational utilities for various common actions performed on the software application, such as downloading a **Portable Document Format** (**PDF**) file, sending an email, and so on.

The test automation framework should also assist engineers in writing tests for components that are otherwise hard to validate. For example, there could be wrapper code written as part of the test automation framework that could be used to call a legacy system as part of an integration test. Another use case would be to support the mocking of external vendor calls within the framework as part of E2E tests.

Test automation should also support tackling bug detection and analysis with minimum human intervention. It is vital to put in place concise logging mechanisms within the framework for swift debugging of failures. Integrations with tools such as **JIRA** and **Bamboo** can be explored to maintain uniform reporting of bugs and issues in an automated manner.

It is the responsibility of the test engineers and the **software development engineers in test** (**SDETs**) to work with the infrastructure team to ensure tests are created and run in the most efficient way within the limitations of a test environment. Test automation will be cost-prohibitive if none of the sub-systems is stable or testable. Test engineers always need to be on the lookout for environment-related errors to establish an uninterrupted automated testing process.

Building a solid test automation framework is only the beginning of the quality journey. There is constant effort involved in maintaining and keeping the framework updated. Let us look at some key factors in maintaining the framework.

Maintaining the framework

One of the chief goals of a test automation framework is to ensure continuous synchronization between the different tools and libraries used. It helps to enable automatic updates of various third-party libraries used in the framework as part of the **continuous integration** (**CI**) pipeline to circumvent manual intervention. Any test tool utilized within the framework should constantly be evaluated for enhancements and security vulnerabilities. Documenting the best practices around handling the framework helps newly onboarded engineers to hit the ground running in less time. Having dedicated SDETs acting as a liaison with tool vendors and users adds stability to the overall test automation process.

It goes without saying the importance of keeping test scripts updated with code changes. SDETs need to stay on top of code changes that render the automation framework inoperable, while the test engineers should continuously look out for code changes that break the test scripts. Test engineers should efficiently balance their time between adding new tests versus updating existing ones to keep the CI pipeline functional. Test automation framework maintenance takes time and sometimes involves significant code changes. It is important to plan and allocate time upfront for the maintenance and enhancements of the framework. It is crucial to spend more time designing and optimizing the framework as this helps in saving time in maintenance down the road.

Next, let us review key factors in handling test data as part of the test automation framework.

Sanitizing test data

Acquiring test data and cleaning it up for sustainable usage in testing is a key task of an automation framework. A lot of cross-team collaboration must happen to build a repository of high-quality test data. Test engineers and SDETs often work with product teams and data analysts to map test scenarios with the datasets that might already be available. Additional queries will have to be run against production databases to cover any gaps identified in this process. Test data thus acquired is static and should be masked to eliminate any **personally identifiable information** (**PII**) for usage within the test automation framework, calling out some of this static data.

A plethora of data-generating libraries are also available for integration with the framework. These libraries can be equipped to handle simple spreadsheets to generate unstructured data based on business logic. Often, there is a need to write custom code to extract, generate, and store test data due to the complexity of software applications. Every automation framework is unique and requires a combination of static and dynamic data generation techniques. It is a good practice to isolate these test data generation libraries for portability and reusability across the organization.

Always try to focus on the quality of test data over quantity. The framework should also include utilities for cleaning up the test data after every test execution to preserve the state of the application. There should be common methods set up to handle test data creation and deletion as part of every test automation framework.

Test data management is a separate domain by itself, and we are just scratching the surface. Let us quickly look at some additional factors in handling and maintaining sanity around test data, as follows:

- Maintaining data integrity throughout its usage in test scripts is essential to the reliability of tests. The automation framework should offer options to clean and reuse test data when and where necessary.

- Guarding the test data for use outside of automation should be avoided at all costs. Maintaining clear communication on which data is being used for test automation keeps the automated tests cleaner.

- Performing data backups of baseline data used for automated testing and using them in the future will help immensely in recreating data across test environments.

- Using data stubs/mocks when dealing with external vendor data or for internal **application programming interfaces** (**APIs**) that may not have **high availability** (**HA**). Typical responses can be saved and replayed by the stubs/mocks wherever necessary in the framework.

- Employing data generators to introduce randomness and to avoid using real-time data within the tests. There are numerous generators available for commonly used data such as addresses and PII.

- Considering all the integration points within the application when implementing utilities that create underlying test data.

- Including **user interface** (**UI**) limitations when designing test data for E2E tests. This is especially useful when using API calls to set up data for UI tests as there might be incompatibilities between how data is handled by different layers of the application stack.

The aspects of the test automation framework that we reviewed so far are mostly agnostic to the hosting environment—that is, on-premise or cloud—but there are certain distinct benefits to embracing the cloud in the overall test automation strategy. Let us look at them in the next section.

Testing in the cloud

Tapping into the advantages of a cloud-based environment is an indispensable aspect of any test automation strategy in the current software engineering landscape. A rightly employed cloud-based testing strategy delivers the dual advantage of faster **time-to-market** (**TTM**) with much lower ownership costs. The biggest advantage of a cloud-based environment is the ability to access resources anytime and anywhere. Geographically diverged teams can scale test environments up and down for their testing needs at will.

Imagine a software engineering team working on developing an application that must be supported across all devices and browsers. Setting up and maintaining test environments that sustain all these configurations can be both tedious and expensive. Cloud-based tools can assist in overcoming this drawback by using preconfigured environments and enabling parallel testing across these platforms. These environments can also be scaled up based on usage. This eliminates the need to purchase and maintain the necessary hardware. Cloud-based tools also provide the on-demand capacity needed for non-functional tests.

Cloud-based environments are foundational pillars of **development-operations** (**DevOps**). With quality built into the DevOps processes, test engineers can spin up test environments and integrate them with several tools as part of their automation framework. With enhanced scalability and flexibility, a cloud-based test automation strategy serves major testing challenges and reduces **go-to-market** (**GTM**) times.

With a good notion of various aspects of the test automation framework, let us next dive into understanding the test pyramid.

Understanding the test pyramid

The test pyramid is a diagrammatic representation of the distribution of different kinds of tests. It acts as a guideline when planning for the test coverage of a software product. It also helps to keep in mind the scope of each test so that it remains within the designed level, thereby increasing the portability and stability of the test. The following diagram summarizes the test automation pyramid:

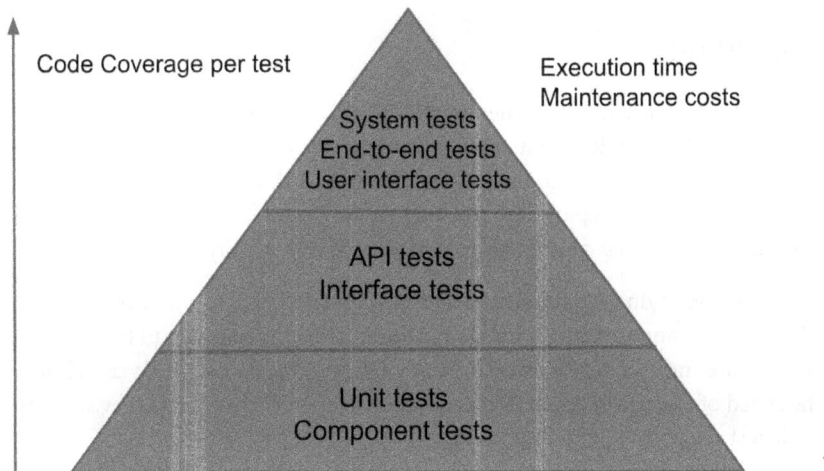

Figure 2.4 – Test automation pyramid

Let us look at each layer of the pyramid in a little more detail.

Unit/component tests

They are tests defined at the lowest level of the system. They are written by the engineer developing the feature, and the involvement of test engineers is minimal (and sometimes zero) in these tests. They form a healthy collection of tests, providing excellent coverage for all the code components. Frameworks used to put these tests in place include xUnit, Rspec, and more. They are completely automated, and these tests should ideally run for every commit to the code repository. External and database calls are stubbed in this layer, which contributes to the increased speed of these tests. Since they are the quickest and the least expensive to write, they come with the biggest ROI among all other types of tests. They also enable swift feedback due to their high frequency of execution.

Certain non-functional tests can also be included as component tests in this layer, with additional setup and partnership with the infrastructure team. Test automation strategy should also comprise techniques to quarantine flaky unit tests. This is usually the bulkiest layer of the pyramid, and any application that can be tested should be included as part of unit/component tests.

The main drawback of unit tests is that they fall short in identifying bugs and errors resulting from the communication of multiple code modules.

Next, let us look at how integration/API tests help with validating communication that extends beyond a module of code.

Integration/API tests

The second layer in the pyramid is where the majority of the application logic is validated. Tests in this layer can be tied directly back to a feature or a combination of features. Ideally, these tests should be executed before merging the code into the `master` branch. If that is not possible, these should be run as soon as the code change is deployed to a testing environment. Since these tests make frequent database or other mocked external calls, their execution time is generally high.

This layer is also apt for verifying API contracts and how APIs communicate with each other. After analyzing production data and setting up inputs for these APIs, the output can be compared to the expected behavior. Care must be taken to exclude any UI-based tests in this layer. While there is a comprise on the speed of execution, these tests tend to offer higher coverage as they span more lines of code in a single test.

Integration/API tests sometimes can be time-consuming and resource-intensive when the link between every API and module must be verified. This is where system tests come in handy. Let us examine and understand the top layer of the pyramid next, comprising system tests.

E2E/System/UI tests

These are tests that exercise the software application through its UI or a series of API calls to validate a business flow. These require a test environment with all the dependencies, associated databases, and mocked vendor calls. These tests are not run as frequently as unit and API tests due to their extensive runtimes. The main goal of these tests is to give product and engineering teams the confidence to deploy the code to production. Since most of the bugs should already be found by the lower-level tests, the frequency of execution of these tests can be adjusted to synchronize with the production release cycles.

These tests also act as acceptance tests for business teams or even the client in some projects. An example would be Selenium tests, which simulate how a real user performs actions on the application. They verify all aspects of the application like a real user would do. Controls must be in place to avoid using real user data from production.

Next, let us examine how to structure the different types of tests for optimal results.

Structuring the test cycles

There are certain standards to keep in mind when structuring these test cycles, as outlined here:

- Automate as many test cases as you need, but no more, and automate them at the lowest level that you can.

- Always try to curb the scope of each test case to one test condition or one business rule. Although this may be difficult in some cases, it goes a long way in keeping your test suite concise and makes debugging much easier.

- It is essential to have clarity on the purpose of each test and avoid dependencies between tests because they quickly increase complexity and maintenance expenses.

- Plan test execution intervals based on test types to maximize bug detection and minimize execution times of the CI pipeline.

- Identify and implement mechanisms to catch and quarantine flaky tests as part of every test execution.

We saw in detail how the test pyramid assists in strategizing test automation. Let us next dive into some common design patterns used in test automation.

Familiarizing ourselves with common design patterns

Design patterns can be viewed as a solution template for addressing commonly occurring issues in software design. Once the underlying software design problem is analyzed and understood, these design patterns can be applied to common problems. Design patterns also help create a shared design language for the software development community. Test automation projects tend to be started in an isolated manner and later scraped due to scalability issues. In the next section, let us try to understand how design patterns help test automation achieve its goals.

Using design patterns in test automation

Test automation is as involved as any other coding undertaking and has its own set of design challenges. But to our rescue, some existing approaches can be applied to solve common design challenges in test automation. Test engineers and SDETs should always be on the lookout for opportunities to optimize test scripts by employing the best design procedures. Some common advantages of using design patterns in test automation include the following:

- Helps with structuring code consistently

- Improves code collaboration

- Promotes reusability of code

- Saves time and effort in addressing common test automation design challenges
- Reduces code maintenance costs

Some common design patterns employed in test automation are noted here:

- **Page Object Model (POM)**
- The factories pattern
- Business layer pattern

Let us now look at each of them in detail.

POM

Object repositories, in general, help keep the objects used in a test script in a central location rather than having them spread across the tests. POM is one of the most used design patterns in test automation; it aids in minimizing duplicate code and makes code maintenance easier. A page object is a class defined to hold elements and methods related to a page on the UI, and this object can be instantiated within the test script. The test can then use these elements and methods to interact with the elements on the page.

Let us imagine a simple web page that serves as an application for various kinds of loans (such as personal loans, quick money loans, and so on). There may be multiple business flows associated with this single web page, and these can be set up as different test cases with distinct outcomes. The test script would be accessing the same UI elements for these flows except when selecting the type of loan to apply for. POM would be a useful design pattern here as the UI elements can be declared within the page object class and utilized in each of the different tests running the business flows. Whenever there is an addition or change to the elements on the UI, the page object class is the only place to be updated.

The following code snippet illustrates the creation of a simple page object class and how the `test_search_title` test uses common elements on the UI from the `Home_Page_Object` page object class to perform its actions:

```
import selenium.webdriver as webdriver
from selenium.webdriver.common.by import By

class WebDriverManagerFactory:
def getWebdriverForBrowser(browserName):
if browserName=='firefox':
return webdriver.Firefox()
elif browserName=='chrome':
return webdriver.Chrome()
elif browserName=='edge':
```

```
return webdriver.Edge()
else:
return 'No match'
```

The `WebDriverManagerFactory` class contains a method to select the driver corresponding to the browser being used, as illustrated in the following code snippet:

```
class Home_Page_Object:
def __init__(self, driver):
self.driver = driver
def load_home_page(self):
self.driver.get("https://www.packtpub.com/")
return self
def load_page(self,url):
self.driver.get(url)
return self
def search_for_title(self, search_text):
self.driver.find_element(By.ID,'search').send_keys(search_text)
search_button=self.driver.find_element(By.XPATH, '//button[@
class="action search"]')
search_button.click()


def tear_down(self):
self.driver.close()
```

The POM paradigm usually has a base class that contains methods for identifying various elements on the page and the actions to be performed on them. The `Home_Page_Object` class here in this example has methods to set up the driver, load the home page, search for titles, and close the driver:

```
 class Test_Script:
def test_1(self):
driver = WebDriverManagerFactory
.getWebdriverForBrowser("chrome")
if driver == 'No match':
raise Exception("No matching browsers found")
pageObject = Home_Page_Object(driver)
pageObject.load_home_page()
pageObject.search_for_title('quality')
```

```
pageObject.tear_down()


def main():
test_executor = Test_Script()
test_executor.test_1()



if __name__ == "__main__":
main()
```

The `test_search_title` method sets up the Chrome driver and the page object to search for the `quality` string.

We will look in further detail about setting up and using POM in *Chapter 5, Test Automation for Web*.

Now, let us investigate how the factory design pattern is helpful in test automation.

The factories pattern

The factory pattern is one of the most used design patterns in test automation and aids in creating and managing test data. The creation and maintenance of test data within a test automation framework could easily get messy, and this approach provides a clean way to create the required objects in the test script, thereby decoupling the specifics of the factory classes from the automation framework. Separating the data logic from the test also helps test engineers keep the code clean, maintainable, and easier to read. This is often achieved in test automation by using pre-built libraries and instantiating objects from classes exposed by the libraries. Test engineers can use the resulting object in their scripts without the need to modify any of the underlying implementations.

A classic example of a factory design pattern in test automation would be how **Selenium WebDriver** gets initialized and passed around in a test. Selenium WebDriver is a framework that enables the execution of cross-browser tests.

The following diagram breaks down how a piece of test script can exercise Selenium WebDriver to make cross-browser calls:

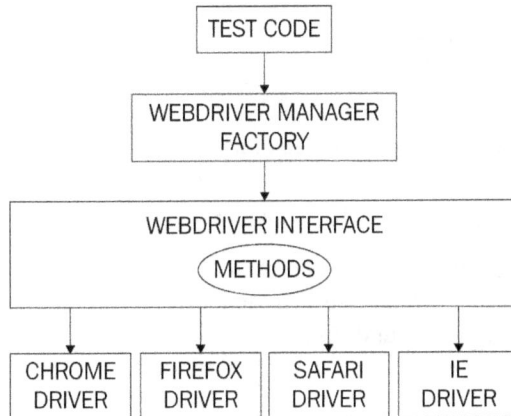

Figure 2.5 – Selenium WebDriver architecture

Figure 2.5 shows how the test code utilizes a factory method to initialize and use the web drivers.

Please refer to the code snippet in the previous section for a simple implementation of the factory pattern. Here, the `WebDriverManagerFactory` class returns an instance of the web driver for the requested browser. The `Test_Search_Choose_Title` class can use the factory method to open a Chrome browser and perform additional validations. Any changes to how drivers are being created are encapsulated from the test script. If Selenium WebDriver supports additional browsers in the future, the factory method will be updated to return the corresponding driver.

Business layer pattern

This is an architectural design pattern where the test code is designed to handle each layer of the application stack separately. The libraries or modules that serve the test script are intentionally broken down into UI, business logic/API, and data handling. This kind of design is immensely helpful when writing E2E tests where there is a constant need to interface with the full stack. For example, the steps involved may be to start with seeding the database with the pre-requisite data, make a series of API calls to execute business flows, and finally validate the UI for correctness. It is critical here to keep the layers separate as this reduces the code maintenance nightmare. Since each of the layers is abstracted, this design pattern promotes reusability. All the business logic is exercised in the API layer, and the UI layer is kept light to enhance the stability of the framework.

Design patterns play a key role in improving the overall test automation process and they should be applied after thoroughly understanding the underlying problem. We need to be wary of the fact that if applied incorrectly, these design patterns could lead to unnecessary complications.

Summary

In this chapter, we dealt primarily with how to come up with a sound test automation strategy. First, we saw some chief objectives that need to be defined to get started with the test automation effort, then we reviewed key aspects in devising a good test automation strategy. We also looked at how the test pyramid helps in formulating a test strategy by breaking down the types of tests. Finally, we surveyed how design patterns can be useful in test automation and learned about some common design patterns.

Now that we have gained solid ground on test automation strategy, in the next chapter, let us look at some useful command-line tools and analyze in detail certain commonly used test automation frameworks and the considerations around them.

Questions

Take a look at the following questions to test your knowledge of the concepts learned in this chapter:

1. Why is a test automation strategy important for a software project?
2. What are the main objectives for test automation?
3. How does the test pyramid help test automation strategy?
4. How do cloud-based tools help in accelerating test automation efforts?
5. What are the advantages of using design patterns in test automation?

3

Common Tools and Frameworks

Every test automation engineer or SDET will have to take advantage of existing tools to build and maintain automation frameworks. For most common use cases, teams could benefit immensely by making use of a plethora of tools and libraries to suit their needs and deliver top results. We introduced test automation and the considerations for devising an effective strategy in the previous two chapters. This chapter will equip you with practical insights and steps to use the **command-line interface** (**CLI**) and **Git**, which are basic skills needed for any test engineer.

We will also touch upon a variety of test automation frameworks and their architecture. This will give you a flavor of different frameworks, their primary purposes, and their benefits. The tools selected for our review here are some of the most commonly used in the market, at this time, and can be applied to different kinds of testing.

The following are the main topics we will be looking at in this chapter:

- The basic tools for every automation engineer
- Common test automation frameworks
- Choosing the right tool/framework

Technical requirements

In this chapter, we will be looking at working examples of the CLI and Git. We will be using the **Terminal** software on macOS for our examples in both sections. Windows users can use the in-built **PowerShell** to execute these commands.

The basic tools for every automation engineer

One of the primary tasks of a test automation engineer is to create, edit, or delete code daily. Test engineers will often also have to interact with the shell of the system under development to tweak their test environments or the underlying test data. In this section, we will be covering a few basic commands that test engineers will need to be able to access the source code and navigate the system under test. This section is a quick refresher for readers who are already experienced in the software engineering space, and can help to build a good foundation for beginners.

Let us start by looking at the CLI.

The CLI

The CLI is a means to interact with the shell of the system under test. A lot of the tasks performed through the graphical user interface can be done through the CLI too. But, the real might of the CLI lies in its ability to programmatically support the simulation of these tasks. Let's try and get familiar with a few basic CLI commands. The CLI commands covered in this section can be run on Terminal software on macOS, or PowerShell on Windows:

- The `ls` command lists all the files and directories in the current folder:

    ```
    →ls
    ```

    The output to the preceding command should be as follows:

    ```
    test.py    test.txt   testing_1 testing_2 testing_3
    testing_4
    →
    ```

- The `cd` command stands for change directory and is used to switch to another directory. The `cd ..` command navigates to the parent directory.

    The syntax is as follows:

    ```
    cd [path_to_directory]
    ```

    The command and output should be as follows:

    ```
    →cd testing_1→ testing_1 cd ..
    ```

- The `mkdir` command creates a new directory under the current directory.

    The syntax is as follows:

    ```
    mkdir [directory_name]
    ```

    The command and output should be as follows:

    ```
    →mkdir testing_5
    →ls
    test.py    test.txt   testing_1 testing_2
    testing_3     testing_4 testing_5
    →
    ```

- The `touch` command creates a new file in the current directory without a preview.

  The syntax is as follows:

  ```
  touch [file_name]
  ```

  The command and output should be as follows:

  ```
  → touch testing.txt
  →ls
  test.py       test.txt     testing.txt testing_1     testing_2
          testing_3     testing_4     testing_5
  ```

> **Note**
> Windows PowerShell users can use `ni` as `touch` is not supported.

- The `cat` command allows the user to view file contents on the CLI.

  The syntax is as follows:

  ```
  cat [file_name]
  ```

  The command and output should be as follows:

  ```
  → cat vim_file
  This is a new file
  →  cli_demo
  ```

So far, we have looked at how to create and modify files. Next, let us look at the commands for deleting files and folders:

- The `rm` command can be used to delete folders and files. Let us look at some specific examples of how to go about this deletion.

  To remove a directory and all the contents under that directory, use the `rm` command with the `-r` option.

  The syntax is as follows:

  ```
  rm -r [directory_name]
  ```

  The command and output should be as follows:

  ```
  → ls
  test.py       test.txt     testing.txt testing_1     testing_2
     testing_3     testing_4     testing_5     vim_file
  → rm -r testing_1
  ```

```
→ls
test.py       test.txt      testing.txt testing_2     testing_3
   testing_4   testing_5    vim_file
→
```

- To delete the file(s), the same rm command can be used followed by the filename.

  The syntax is as follows:

  ```
  rm [file_name]
  ```

  The command and output should be as follows:

  ```
  →  cli_demo ls
  test.py       test.txt      testing.txt testing_2     testing_3
     testing_4   testing_5    vim_file
  →  cli_demo rm test.txt
  →  cli_demo ls
  test.py       testing.txt testing_2     testing_3
     testing_4   testing_5    vim_file
  →  cli_demo
  ```

Next, let us quickly look at **Vim**, which is a commonly used file-handling tool for the CLI.

Working with Vim

Vim is an in-built editor that allows you to modify the contents of a file. Vim aims to increase efficiency when editing code via the CLI and is supported across all major platforms, such as macOS, Windows, and Linux. Vim also supports creating custom keyboard shortcuts based on your typing needs. Let's look at a basic example of editing and saving a file. This editor supports a wide range of commands and can be referenced at http://vimdoc.sourceforge.net/. To edit and save a file, you need to do the following:

1. To execute the editor, the user has to type vi, followed by a space and the filename:

   ```
   workspace vi test.txt
   ```

2. Then, type i to switch to INSERT mode and type in the contents of the file.

3. Press the *Esc* key to quit INSERT mode.

4. Next, type :wq to save and exit. This command is a combination of :w to write the contents of the file to the disk and q to quit the file.

5. Then, press i to enter INSERT mode and type the required contents in the file.

The CLI commands we have looked at so far should serve as a good starting point for new users. Now, let us familiarize ourselves with flags in the CLI.

Flags in the CLI

Flags are add-ons to enhance the usage of a command in the CLI. For example, the −l flag can be applied to the ls command to alter the displayed list of files and folders in a long format.

The syntax is as follows:

```
ls -l
```

The command and output should be as follows:

```
→ls -l
total 8
-rw-r--r--  1 packt  staff   0 Jun 25 09:33 test.py
-rw-r--r--  1 packt  staff   0 Aug 14 18:14 testing.txt
drwxr-xr-x  3 packt  staff  96 Jun 26 10:20 testing_2
drwxr-xr-x  2 packt  staff  64 Jun 25 09:32 testing_3
drwxr-xr-x  2 packt  staff  64 Jun 25 23:16 testing_4
drwxr-xr-x  2 packt  staff  64 Aug 14 18:13 testing_5
-rw-r--r--  1 packt  staff  19 Aug 14 18:26 vim_file
→
```

There are thousands of flags that can be attached to various CLI commands, and it is impossible to know all of them. This is where the man command comes in handy. man can be used with any CLI command, and it gives all the options and an associated description for each command. There are usually multiple pages of help content and you are encouraged to browse through them.

For example, to learn all the information associated with the ls commands, you just have to run the following command:

```
man ls
```

There are a few tips/tricks to keep in mind regarding CLI usage, such as the following:

- All the CLI commands are case sensitive
- The pwd command lists the current working directory
- The clear command clears the contents on the current shell window
- The *up/down* arrow keys can be used to navigate through the history of the CLI commands
- The *Tab* key can be used to get autocomplete suggestions based on the string typed so far

- The cd – and cd ~ commands can be used to navigate to the last working directory and home directory, respectively

- Multiple CLI commands can be run in a single line using the ; separator

The power of shell scripting

The ultimate utility of the CLI lies in writing automated scripts that perform repeatable tasks. Shell scripting can be used to achieve this and can save you a great deal of time. Users are encouraged to refer to the full documentation at https://www.gnu.org/software/bash/manual/bash.html to learn more about commands and their syntax. To understand the power of the CLI, let us look at an example of a shell script in this section. This script creates a folder, named test_folder, and then creates a text file, named test_file, within it. The script then uses the curl command to download a web resource that is passed as an argument and stores its output in test_file.txt. Now, $1 refers to the first argument used when invoking this file for execution. -o is used to override the contents of the file. Then, it reads the file using the cat command and stores it in a variable named file_content. Finally, this file is removed using the rm command:

```
#!/bin/bash
mkdir test_folder
cd test_folder
touch test_file.txt
curl $1  -o test_file.txt
file_content=`cat test_file.txt`
echo $file_content
rm test_file.txt
```

This script can be executed using the bash sample_bash_script.sh https://www.packt.com/ command, where sample_bash_script.sh is the name of the file. Please note that the web resource here can be downloaded at https://www.packt.com/ and that it is being passed as the first argument to the script.

We have just gotten a bird's eye view of the CLI, and I strongly encourage you to dive deeper into CLI commands to increase your proficiency. Some major advantages of using the CLI include the following:

- **Speed and security**: CLI commands are faster and more secure to use than the corresponding actions being done through the graphical user interface.

- **Scripting on the CLI**: The CLI lets users write scripts to perform repetitive actions by combining them into a single script file. This is much more stable and efficient than a script run on a graphical user interface.

- **Resource efficient**: CLI commands use much fewer system resources and therefore provide better stability.

Now that we have familiarized ourselves with the CLI, let us look at another tool that is an absolute necessity for the maintenance of a software project of any size.

Git

Git is a modern distributed version control system that allows tracking changes to the source code and is a versatile tool to enable collaboration in the engineering team. Git primarily helps in synchronizing contributions to source code by various members of the team, by keeping track of the progress over time.

Every software application is broken down into code repositories and production code is stored on a branch called master on the repository. When an engineer is ready to begin working on a feature, they can clone the repository locally and create a new branch to make their changes. After the code changes are complete, the engineer creates a pull request that is then peer-reviewed and approved. This is when they are merged into the master branch. Subsequently, the changes are deployed to the staging and production environments. There are various hosting services, such as GitHub, that provide a user interface to maintain, track, and coordinate contributions to the code repositories. Now, let us look at some of the common Git commands that test engineers might have to use frequently:

- `git --version` shows the version of Git software installed on the machine:

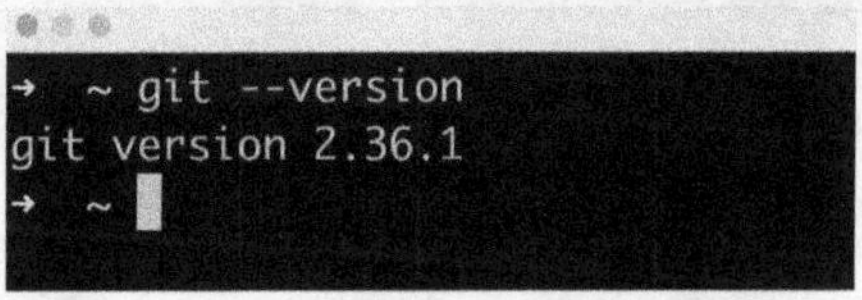

```
→  ~ git --version
git version 2.36.1
→  ~ ▮
```

Figure 3.1 – git version

- `git init` initializes the project folder into a GitHub repository:

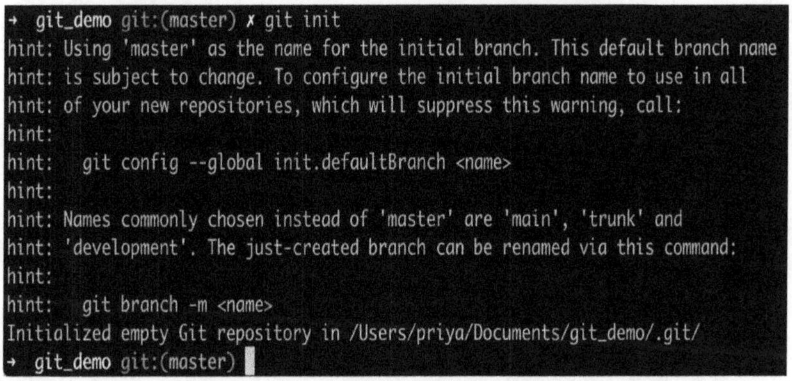

```
→ git_demo git:(master) ✗ git init
hint: Using 'master' as the name for the initial branch. This default branch name
hint: is subject to change. To configure the initial branch name to use in all
hint: of your new repositories, which will suppress this warning, call:
hint:
hint:   git config --global init.defaultBranch <name>
hint:
hint: Names commonly chosen instead of 'master' are 'main', 'trunk' and
hint: 'development'. The just-created branch can be renamed via this command:
hint:
hint:   git branch -m <name>
Initialized empty Git repository in /Users/priya/Documents/git_demo/.git/
→ git_demo git:(master) ▮
```

Figure 3.2 – git init

- `git clone [repository_URL]` creates a local copy of the remote repository:

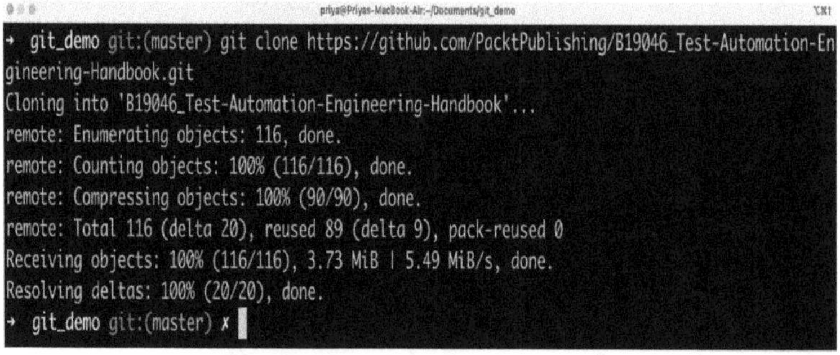

Figure 3.3 – git clone

- `git push` pushes all of the committed local changes to the remote GitHub repository:

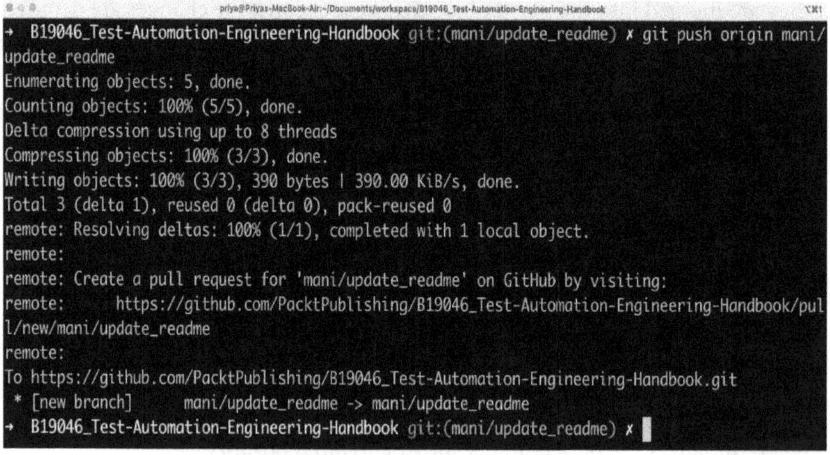

Figure 3.4 – git push

- `git pull` pulls all the latest code from the remote branch and merges them with the local branch:

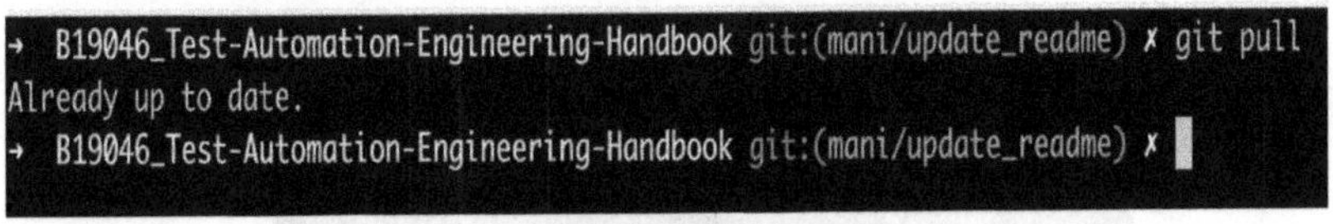

```
→ B19046_Test-Automation-Engineering-Handbook git:(mani/update_readme) ✗ git pull
Already up to date.
→ B19046_Test-Automation-Engineering-Handbook git:(mani/update_readme) ✗ ▮
```

Figure 3.5 – git pull

- `git log` lists the entire commit history:

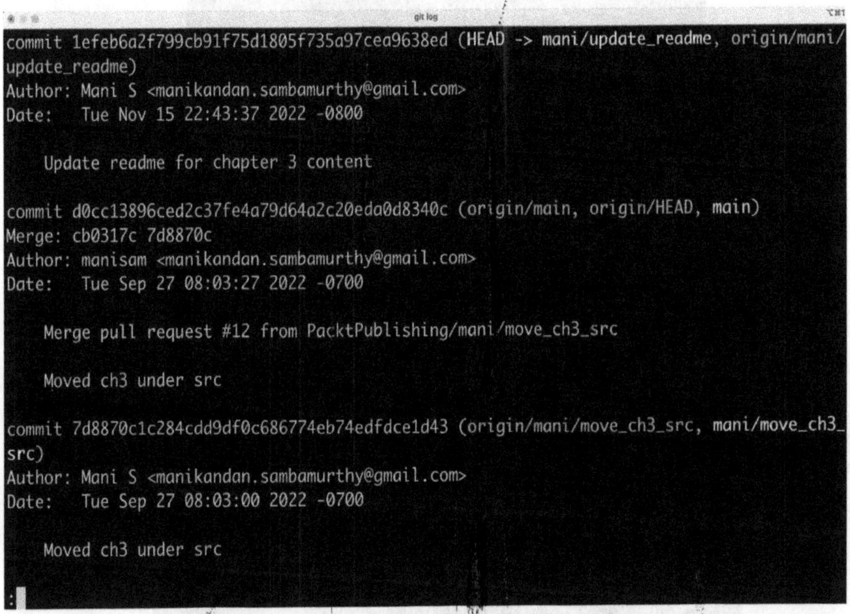

```
• • •                                         git log                                          ⌥⌘1
commit 1efeb6a2f799cb91f75d1805f735a97cea9638ed (HEAD -> mani/update_readme, origin/mani/
update_readme)
Author: Mani S <manikandan.sambamurthy@gmail.com>
Date:   Tue Nov 15 22:43:37 2022 -0800

    Update readme for chapter 3 content

commit d0cc13896ced2c37fe4a79d64a2c20eda0d8340c (origin/main, origin/HEAD, main)
Merge: cb0317c 7d8870c
Author: manisam <manikandan.sambamurthy@gmail.com>
Date:   Tue Sep 27 08:03:27 2022 -0700

    Merge pull request #12 from PacktPublishing/mani/move_ch3_src

    Moved ch3 under src

commit 7d8870c1c284cdd9df0c686774eb74edfdce1d43 (origin/mani/move_ch3_src, mani/move_ch3_
src)
Author: Mani S <manikandan.sambamurthy@gmail.com>
Date:   Tue Sep 27 08:03:00 2022 -0700

    Moved ch3 under src

:
```

Figure 3.6 – git log

- `git branch [branch_name]` creates a new branch in the local Git repository:

```
• • •              priya@Priyas-MacBook-Air-/Documents/workspace/B18046_Test-Automation-Engineering-Handbook              ⌥⌘1
→ B19046_Test-Automation-Engineering-Handbook git:(mani/update_readme) ✗ git branch mani
/git-demo-branch
→ B19046_Test-Automation-Engineering-Handbook git:(mani/update_readme) ✗ ▮
```

Figure 3.7 – git branch [branch_name]

- `git branch` lists all the local branches created so far. `*` indicates the branch that is currently checked out:

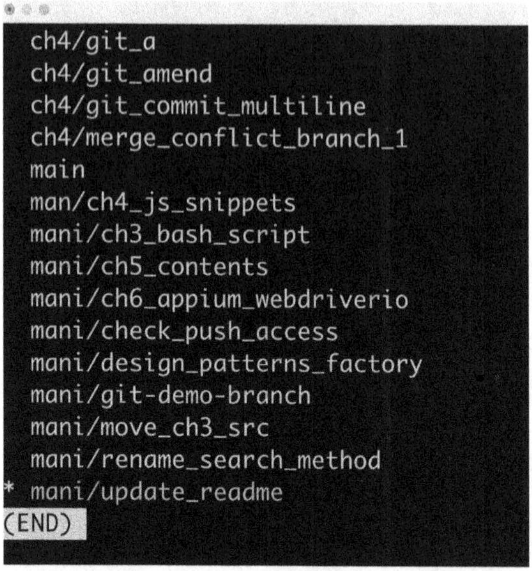

Figure 3.8 – git branch

- `git branch -a` lists all the local and remote branches created so far:

Figure 3.9 – git branch -a

- `git checkout [branch_name]` switches between local Git branches:

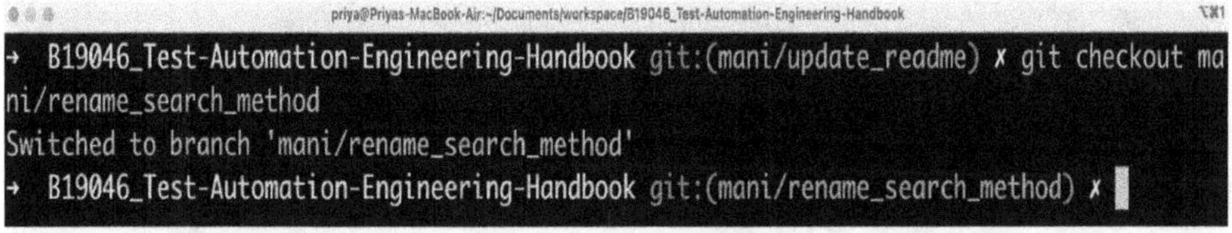

Figure 3.10 – git checkout

- `git status` displays the modified files and folders in the current project repository:

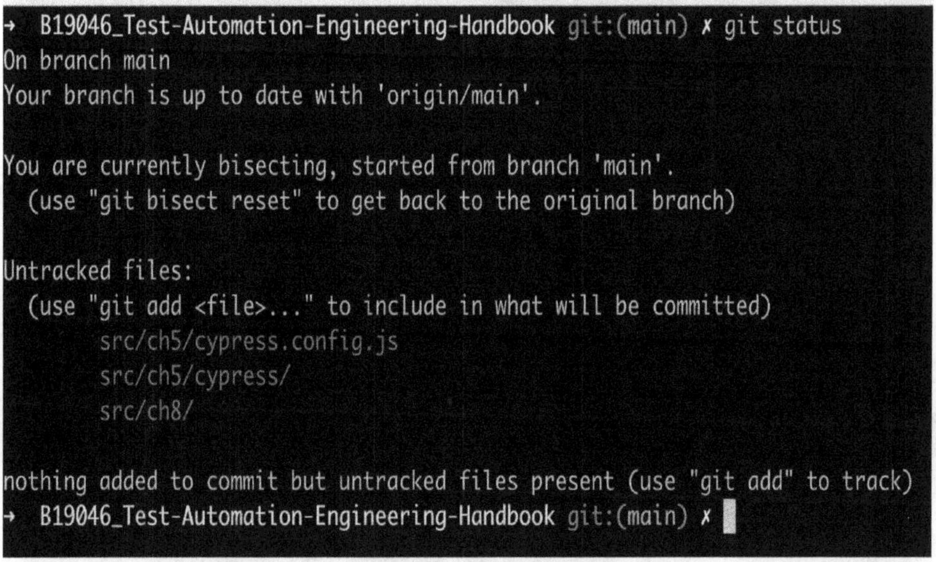

Figure 3.11 – git status

- `git diff` shows the difference between files in the staging area and the working tree:

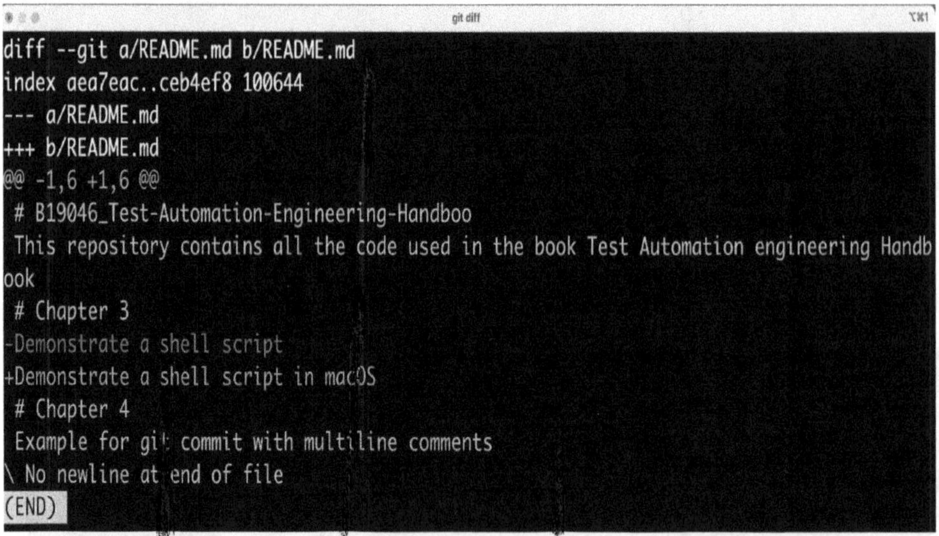

```
diff --git a/README.md b/README.md
index aea7eac..ceb4ef8 100644
--- a/README.md
+++ b/README.md
@@ -1,6 +1,6 @@
 # B19046_Test-Automation-Engineering-Handboo
 This repository contains all the code used in the book Test Automation engineering Handb
ook
 # Chapter 3
-Demonstrate a shell script
+Demonstrate a shell script in macOS
 # Chapter 4
 Example for git commit with multiline comments
\ No newline at end of file
(END)
```

Figure 3.12 – git diff

- `git add .` adds all the modified files to the Git staging area:

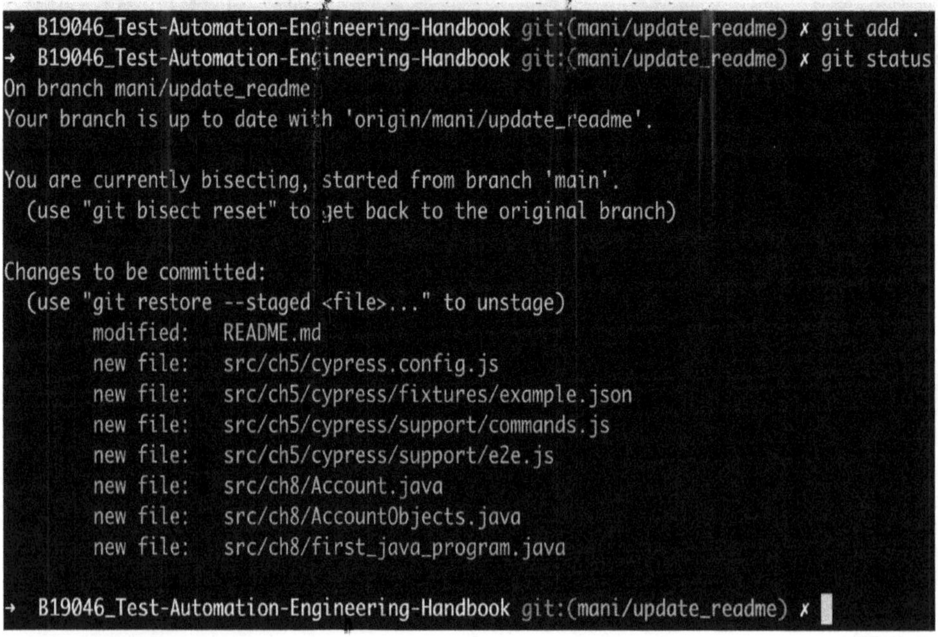

```
→  B19046_Test-Automation-Engineering-Handbook git:(mani/update_readme) x git add .
→  B19046_Test-Automation-Engineering-Handbook git:(mani/update_readme) x git status
On branch mani/update_readme
Your branch is up to date with 'origin/mani/update_readme'.

You are currently bisecting, started from branch 'main'.
  (use "git bisect reset" to get back to the original branch)

Changes to be committed:
  (use "git restore --staged <file>..." to unstage)
        modified:   README.md
        new file:   src/ch5/cypress.config.js
        new file:   src/ch5/cypress/fixtures/example.json
        new file:   src/ch5/cypress/support/commands.js
        new file:   src/ch5/cypress/support/e2e.js
        new file:   src/ch8/Account.java
        new file:   src/ch8/AccountObjects.java
        new file:   src/ch8/first_java_program.java

→  B19046_Test-Automation-Engineering-Handbook git:(mani/update_readme) x
```

Figure 3.13 – git add

- `git commit -m "commit description"` saves the changes to the local repository with the provided description:

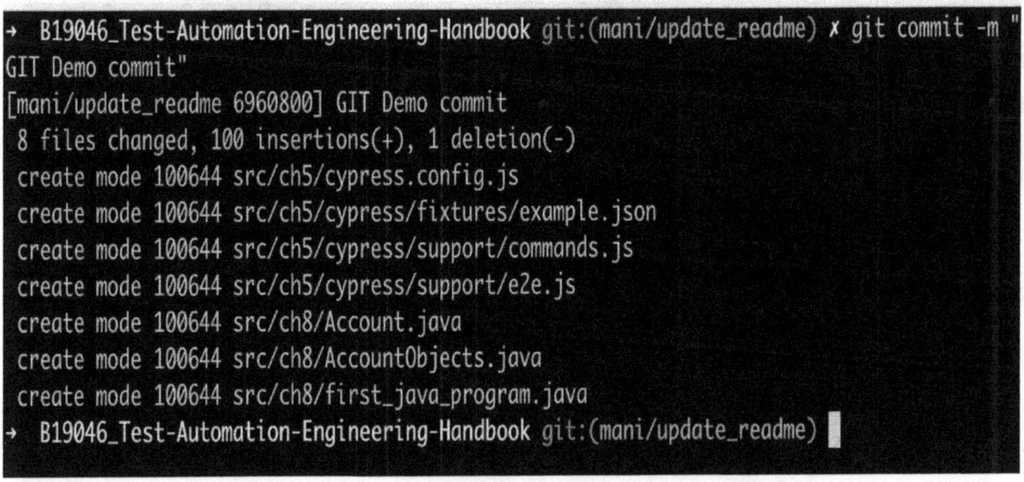

```
→ B19046_Test-Automation-Engineering-Handbook git:(mani/update_readme) ✗ git commit -m "
GIT Demo commit"
[mani/update_readme 6960800] GIT Demo commit
 8 files changed, 100 insertions(+), 1 deletion(-)
 create mode 100644 src/ch5/cypress.config.js
 create mode 100644 src/ch5/cypress/fixtures/example.json
 create mode 100644 src/ch5/cypress/support/commands.js
 create mode 100644 src/ch5/cypress/support/e2e.js
 create mode 100644 src/ch8/Account.java
 create mode 100644 src/ch8/AccountObjects.java
 create mode 100644 src/ch8/first_java_program.java
→ B19046_Test-Automation-Engineering-Handbook git:(mani/update_readme)
```

Figure 3.14 – git commit

- `git branch -D [branch_name]` force deletes the specified local branch.
- `git stash` temporarily removes the changes on the local branch. Use `git stash pop` to apply the changes back onto the local branch:

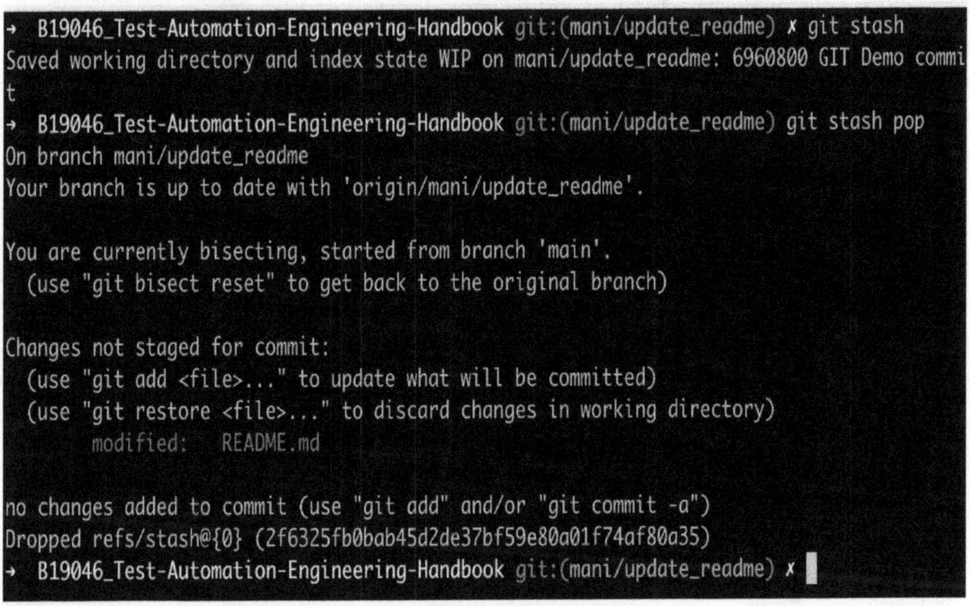

```
→ B19046_Test-Automation-Engineering-Handbook git:(mani/update_readme) ✗ git stash
Saved working directory and index state WIP on mani/update_readme: 6960800 GIT Demo commi
t
→ B19046_Test-Automation-Engineering-Handbook git:(mani/update_readme) git stash pop
On branch mani/update_readme
Your branch is up to date with 'origin/mani/update_readme'.

You are currently bisecting, started from branch 'main'.
  (use "git bisect reset" to get back to the original branch)

Changes not staged for commit:
  (use "git add <file>..." to update what will be committed)
  (use "git restore <file>..." to discard changes in working directory)
        modified:   README.md

no changes added to commit (use "git add" and/or "git commit -a")
Dropped refs/stash@{0} (2f6325fb0bab45d2de37bf59e80a01f74af80a35)
→ B19046_Test-Automation-Engineering-Handbook git:(mani/update_readme) ✗
```

Figure 3.15 – git stash

- `git remote -v` gives the name, as well as the URL, of the remote repository:

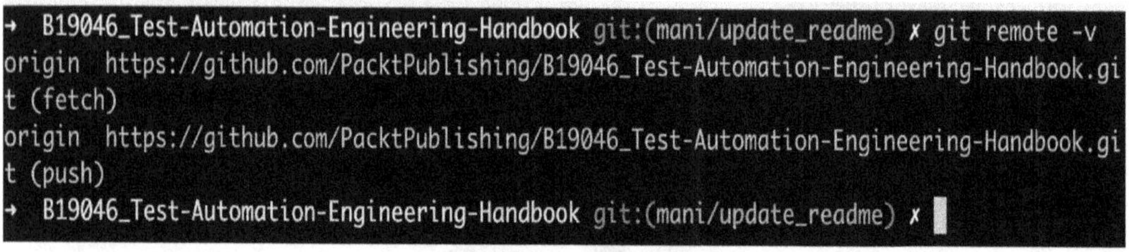

```
→  B19046_Test-Automation-Engineering-Handbook git:(mani/update_readme) x git remote -v
origin  https://github.com/PacktPublishing/B19046_Test-Automation-Engineering-Handbook.gi
t (fetch)
origin  https://github.com/PacktPublishing/B19046_Test-Automation-Engineering-Handbook.gi
t (push)
→  B19046_Test-Automation-Engineering-Handbook git:(mani/update_readme) x ▌
```

Figure 3.16 – git remote

This overview provides you with a healthy introduction to Git and its most commonly used commands. You can explore additional commands and their usage here at `https://git-scm.com/docs/git`. Next, let us dive into some of the most commonly used test automation frameworks.

Common test automation frameworks

Test automation frameworks are a collection of tools and processes that help standardize the test automation effort. They act as technical guidelines for any test implementation. We touched upon the importance of test automation frameworks in *Chapter 2, Test Automation Strategy*, in devising a sound test strategy. In this section, we will look at a selection of frameworks that assist in diverse kinds of testing. For each of these frameworks, we will try to understand their high-level architecture and the primary purpose they serve. In the subsequent chapters of this book, we will implement automated test cases using most of these frameworks. You should note that there is a whole array of effective test automation tools available on the market to serve specific organizational needs. For example, Playwright and Puppeteer are often considered to be modern alternatives for Selenium. The goal of this section is to equip you with knowledge about the basic architecture of some of these tools so that you can explore the other options further in detail.

To assist with our understanding of these frameworks, let us use a hypothetical banking loan application with frontend built-in ReactJS, and a backend comprising REST APIs. Let us ignore the database calls and other external integrations for simplicity. *Figure 3.17* depicts this application with three pages in the frontend of applying for a loan, managing existing loans, and making payments. The backend comprises multiple REST APIs to support the frontend transactions. Let us also assume that Bank XYZ has mobile applications supported in iOS and Android platforms for these transactions. Let us look at how these frameworks can be used to test these applications in the upcoming sections.

BANK XYZ

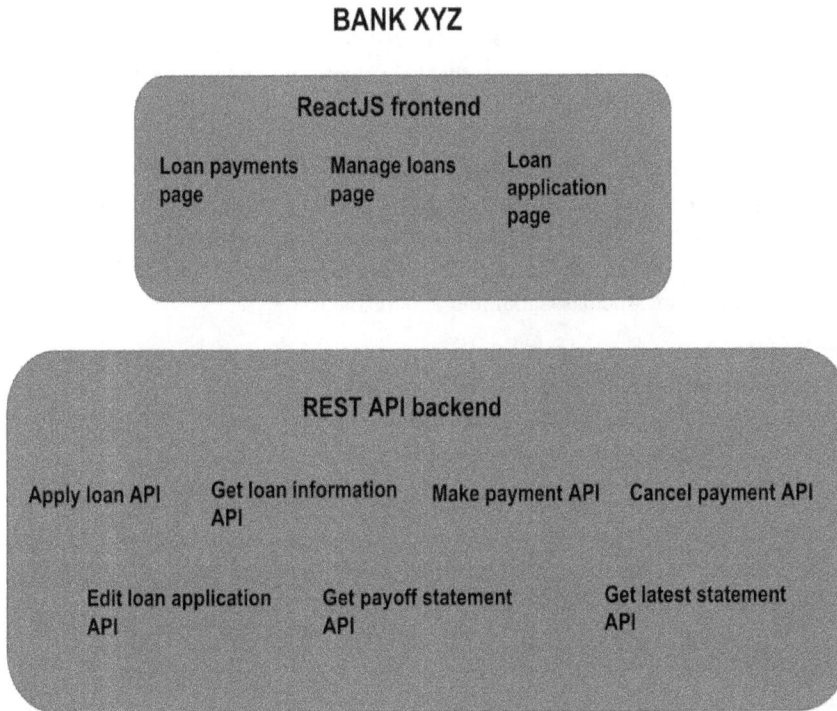

Figure 3.17 – Example loan application

Let us jump right into exploring Selenium.

Selenium

Selenium is an open source tool that provides an interface to automate user actions on web browsers. Imagine a manual tester that must test the login page of an application. There are a variety of scenarios that can be run such as validating various kinds of inputs, empty fields, and incorrect entries. This must be done across multiple browsers such as Chrome, Firefox, and Safari. Also, whenever there is a code change that impacts the login page, these tests must be repeated. Even though this is a simple scenario, in the real world, the efforts involved to test and keep retesting the changes can compound swiftly. When we apply this logic to testing the business flows of a complex enterprise application, modern software engineering teams simply do not have the capacity to scale the manual testing. This is where Selenium can turbocharge browser testing, helping software engineers and product teams tremendously.

Components of Selenium

At a high level, Selenium has three components that help create and orchestrate the tests. Let us look at each one of the following components:

- **WebDriver**: WebDriver is a critical component of the Selenium ecosystem that is responsible for executing the tests by invoking browser-specific drivers. WebDriver comes with an API that uses bindings to support various programming languages, such as Python, Ruby, Java, and C#. WebDriver hosts the API and bindings as a common library. WebDriver also includes support for various integrations, such as Cucumber and TestNG.

- **Grid**: Grid is a server that helps minimize test runtime by routing and balancing the test commands across multiple remote browser instances. It consists of a hub and a node, which handle the requests from the WebDriver and execute them on a remote WebDriver on a different device. Grid can be configured to dynamically scale to suit the test runs.

- **IDE**: The IDE is a record and playback tool that comes as a plugin on Chrome and Firefox. Users can enable the plugin on the browser and record transactions. The IDE internally generates the code in the supported programming languages based on the setting. This code is usually not reusable, and hence is not advised for bigger projects. This works well for quick checks.

In the next section, we will explore the high-level architecture of Selenium.

High-level architecture

We looked at how WebDriver orchestrates the call to execute tests on the browser in *Chapter 2*, specifically the *The Factories pattern* section. Here, we look at the illustration again in *Figure 3.18*. The test script initializes WebDriver, which sends the commands to the local browser instance. The driver executes the commands on the browser instance.

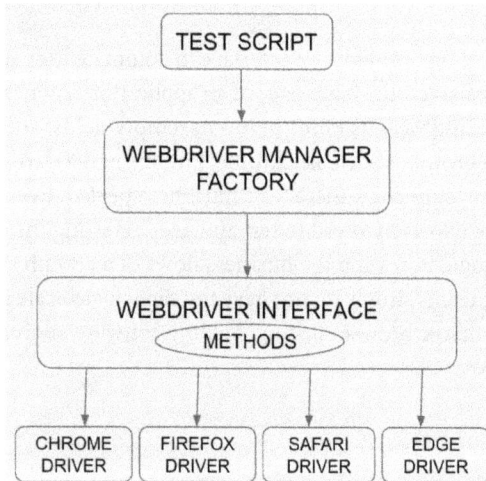

Figure 3.18 – Visual representation of how a test script utilizes a
factory method to initialize and use the drivers

In cases where parallel test runs are required, the hub orchestrates the requests from a client and spreads them across multiple devices containing remote web drivers, as shown in *Figure 3.19*. A remote WebDriver instance has two internal components: a client running the test script, and a server to run the test on a remote device.

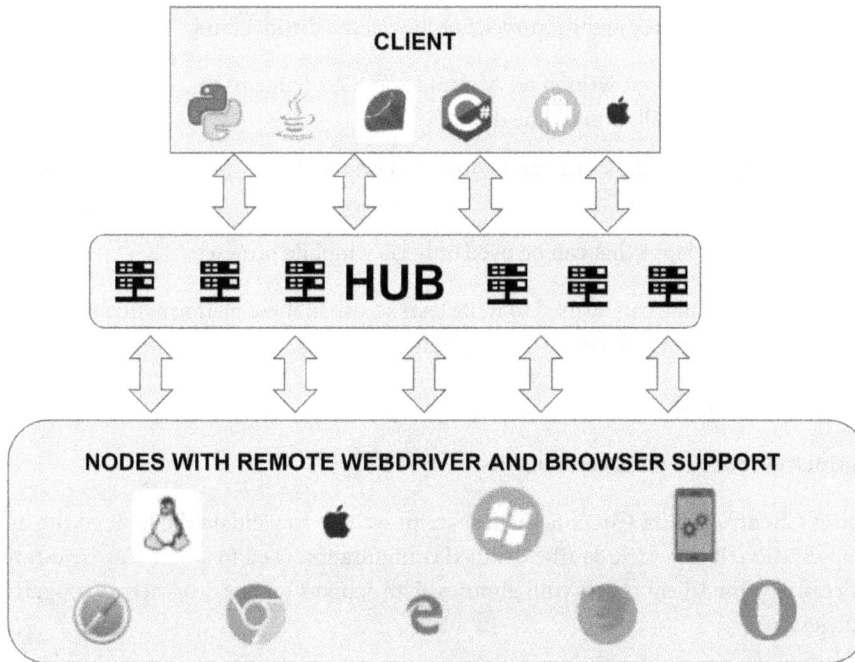

Figure 3.19 – Components of Selenium Grid

Selenium is very flexible in its setup, and hence it's very popular among test and software engineers. Most browsers today have built-in support for Selenium testing, making it a very handy tool. Like any other open source tool, Selenium depends on community contributions for support through third-party libraries. Since it was open-sourced more than a decade ago, there have been continuous improvements done by users. The Selenium project can be found at `https://github.com/SeleniumHQ`.

Now, for the testing of our hypothetical banking application, Selenium can primarily be used to verify frontend aspects. We should be creating page object models for each of the pages and capturing UI elements using appropriate selectors. Then, we need to create a script to load the URL of the application and verify the flow along with the interaction of these elements. Scripts should also be set up to create the drivers and execute them on various browsers.

Even though Selenium is multi-faceted in browser automation, it lacks depth in supporting mobile-native applications. In the next section, let us look at Appium, which is a specialized tool for automating mobile-native applications.

Appium

Appium is an open source tool used for automating tests on platforms such as Android and iOS. It also supports mobile browser automation with extensive community support. Appium provides libraries backing a variety of programming languages, such as Java, Python, PHP, C#, JavaScript, and Ruby. Before diving into Appium's architecture, let us look at the distinct kinds of mobile applications:

- **Native apps**: These are apps written on Android and iOS platforms that can be interacted with only after complete installation on the device

- **Hybrid apps**: These are apps that can be installed on the device, as well as accessed through a browser URL

- **Web apps**: These are apps that can be used only via a mobile browser

Appium offers a single API that can be used to write tests across all these platforms and greatly enables the reuse of test code. Now, let us look at what Appium comprises.

Components of Appium

Appium is primarily made of three components:

- **Appium Client**: Appium Client is the test script written to validate the logic of the mobile/web application. It also includes the detailed configuration used to access the various mobile platforms. Appium Client comes with libraries, with support for a wide variety of programming languages.

- **Appium Server**: Appium Server is the component that interacts with the client and transfers the commands to the end device. It is written in Node.js and parses the request from the client into a JSON format and relays them to the devices or an emulator.

- **Device/emulator**: These are mobile devices or emulators on which the test code sent by the Appium client is executed by the server.

Next, let us look at the architecture of the inner workings of Appium.

High-level architecture

As seen in the previous section, Appium is built on a client-server architecture. As shown in *Figure 3.20*, it comprises an HTTP server written on **Node.js**. Appium Client communicates with the server using RESTful APIs through **MOBILE JSON WIRE PROTOCOL**.

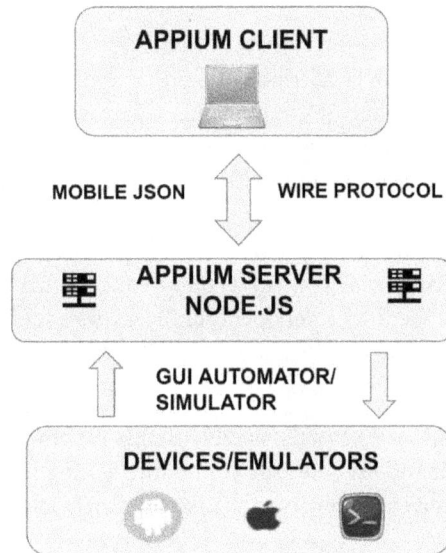

Figure 3.20 – Appium architecture/components

It is vital to know that Appium uses Selenium WebDriver commands to interact with the mobile interface. Appium Client libraries convert the test code into REST API requests and establish a session with the server. These requests are then forwarded by the server to the devices/emulators. Appium Client and the server use this session ID as a reference for their communication. Apple and Android devices have their own APIs to expose the UI elements on their devices. Appium interacts with these elements by sending JSON commands to these APIs.

For testing our hypothetical banking application, we would first set up the Appium server and add platform-specific configuration to support iOS and Android applications. We would also set up virtual devices with an appropriate version of the OS on emulators. With both the emulators and the Appium server set up, we can create and execute test scripts to validate the UI elements from the mobile application. We could also perform a similar configuration for physical mobile devices and run the tests on them.

This is the high-level architecture of the Appium tool. In *Chapter 6, Test Automation for Mobile*, we will look at a test case's implementation using Appium. In the meantime, you can refer to Appium's official documentation at `https://appium.io/docs/en/about-appium/intro/`. Now let us turn our attention to Cypress, which is another effective web automation tool.

Cypress

Cypress is an open source tool written in JavaScript and is primarily used for web automation testing. The main difference between Cypress and Selenium is that Cypress works directly on the browser, whereas Selenium interacts via drivers. Cypress also has the capability to run tests across multiple browsers. It can detect the browsers installed on a particular machine and display them as an option for test execution. Cypress comes in-built with the ability to access both frontend and backend parts of the application. This makes it a great E2E testing tool and is deemed a next-generation test automation tool for the modern web. Cypress is especially easy to set up and quickly functional in JavaScript-based environments. Now, let us look at the chief components of the Cypress tool.

Components of Cypress

Since Cypress is a Node.js application, it has a remarkably simple yet powerful architecture. Cypress sits directly on the browser itself and executes the commands taking full control of the browser automation. The components involved are few, but they are amazingly effective:

- **Cypress Node.js server**: This is the server on which Cypress runs and communicates with the test runner. This server constantly synchronizes the tasks between the test runner and itself, thereby enabling real-time responses to the application events.

- **Cypress test runner**: Cypress comes with an interactive test runner that shows the execution of commands as they run against the application.

Now, let us dig deeper into the architecture and inner workings of the Cypress tool.

High-level architecture

As mentioned in the previous sections, Cypress comprises a Node.js server that sits directly on the browser to execute the test commands. The Node.js server and Cypress constantly interact to keep total control of the frontend and backend aspects of the application under test. Cypress not only accesses the DOM and other browser objects but also monitors the networking layer. This helps it dictate actions both within and outside the browser. *Figure 3.21* provides a diagrammatic representation of this architecture:

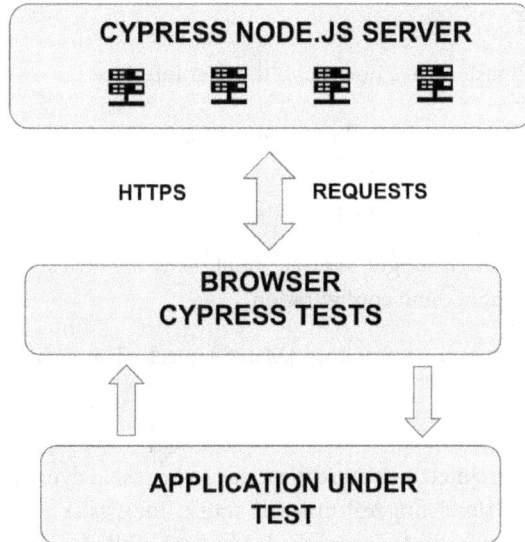

Figure 3.21 – Cypress architecture/components

Since the browser directly runs the Cypress tests, the test's behavior can be modified dynamically by accessing the network traffic. For further reading on Cypress, you can refer to the documentation at `https://docs.cypress.io`. We will implement a test script using the Cypress tool in *Chapter 5, Test Automation for the Web*.

Cypress can be effectively used to set up E2E test scenarios for our hypothetical banking application. We could follow a similar approach as that of Selenium for setting up page objects to verify interaction with the frontend aspects of the application. Additionally, we could intercept the backend API calls and add validations to them. In this case, we would have test scripts covering multiple E2E flows, verifying both UI elements and the API responses involved in that flow.

Next, let us look at a performance testing tool.

JMeter

JMeter is an open source performance testing tool that simulates user requests to a server and then computes and stores the response statistics from the server. It comes with an intuitive user interface that assists in designing various parameters of the tests. It is a versatile tool that can be configured to run in a multithreaded fashion on Linux and Windows machines. It also supports a wide variety of protocols, such as REST, HTTPS, SOAP, Mail/SMTP, and JDBC. Let us look at the different components of the JMeter tool.

Components of JMeter

JMeter primarily comprises master/slave nodes and the user interface:

- **JMeter master**: This is a core component that configures the workers and integrates all the test results
- **JMeter slave nodes**: Distributes virtual machines used to run the tests
- **User interface**: This is a lightweight interface built using the Java swing component that can be used for test maintenance and configuration

Now that we are aware of the components within JMeter, let us look at its high-level architecture.

High-level architecture

JMeter uses the **master-slave** architecture where the tests can be scaled dynamically by increasing the number of slave nodes, thereby simulating real-time web traffic. The master node acts as an orchestrator where the tests are triggered and helps to spread the load across multiple nodes. Tests can be invoked via the CLI or the UI. *Figure 3.22* illustrates the inner workings of the JMeter architecture:

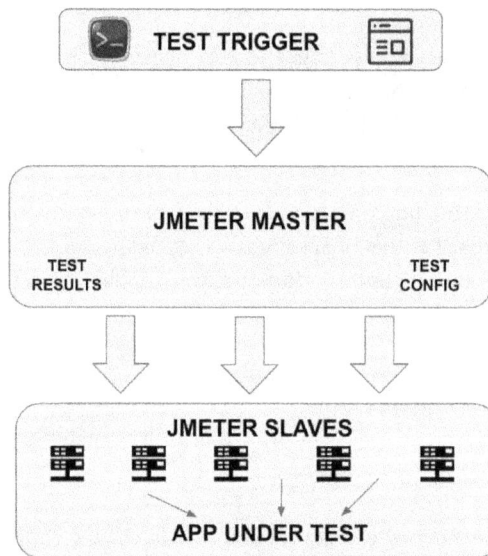

Figure 3.22 – JMeter architecture/components

By sampling many threads simultaneously, JMeter effectively helps determine the performance of web applications. The official documentation for JMeter can be found at `https://jmeter.apache.org/usermanual/index.html`. We will look at a detailed implementation of a performance test in *Chapter 8, Test Automation for Performance*.

In the case of our banking application, we would initially look at setting up performance tests for each of the APIs. Every API should conform to an SLA in terms of the maximum number of concurrent users and response times. JMeter can be used to set up and verify that. Then, we move on to performance test the response times on the UI screens. We normally start with a minimal number of concurrent users and gradually throttle that number to observe the response times of the web pages. We could also stress test both the APIs and the UI pages by increasing the concurrent users to a point where it breaks the application.

Next, let us look at an accessibility testing tool, AXE.

AXE

Axe is a lightweight browser-based accessibility testing tool that is available as an extension. It checks a web page against pre-defined accessibility rules and generates a report of compliances/violations. It also provides options to check parts of a page and generate insights based on the generated report. There are additional libraries available to check compliance against audits and continuously monitor the accessibility status of the web page. You can further explore the capabilities of this tool by referring to `https://www.deque.com/axe/core-documentation/api-documentation/`.

For testing our hypothetical banking application, we would be adding the AXE tool plugin to our browser to get real-time feedback on the level of accessibility support. We could also explore implementing their linters within the application code bases to automatically review every merge.

We have looked at a sample of test automation tools and their high-level architecture so far in this chapter. *Table 3.1* summarizes their primary usage and support:

Tool	Popularly used for	Applications tested	Supported platforms	Supported programming languages
Selenium	Web browser automation	Web, mobile (with external integrations)	Windows/ macOS/Linux	JavaScript, Java, Python, C#, PHP, Ruby, Perl
Appium	Native and hybrid mobile application automation	Mobile	iOS, Android, macOS, Windows	JavaScript, Java, Python, C#, PHP, Ruby
Cypress	E2E testing for web applications	Web	Windows/ macOS/Linux	JavaScript
JMeter	Performance testing of web applications	Web	Windows/ macOS/Linux	Java, Groovy script
AXE	Accessibility testing and associated compliance	Web, mobile	Windows, macOS, iOS, Android	JavaScript, Java, Python, C#, PHP, Ruby

A common question that lingers in the minds of testing teams is how to pick the right tool for the task at hand. In the next section, we will look at some standard considerations in picking the right tool for the various kinds of testing.

Choosing the right tool/framework

In the previous chapter, *Test Automation Strategy*, we pondered over the selection and training needed to implement a test automation tool. In this section, we will reiterate some of those points, and additionally, we will break down the tool selection criteria by the types of testing. Every testing endeavor is unique and demands its own set of tools and techniques. The team composition and the overall maturity of the engineering organization also play a vital role in delivering the ROI on test automation tools. Let us now look at some tool selection guidelines for distinct types of testing.

Selecting a performance testing tool

Performance testing, as an activity, focuses on identifying and documenting the benchmarks for non-functional attributes of a software application. Even though the primary basis for a performance tool selection is cost, skills and the testing approach are also critical. Since every performance testing effort requires concurrent user load generation, there is a license involved on top of the open source aspect of the tool. Let us now review the various performance tool requirements:

- The ability to scale and coordinate concurrent users

- The ability to monitor and record performance metrics

- The ability to provide a user-friendly way to view the performance test results

- Licensing costs and restrictions

- The ability to generate user load across different stacks of the application

- The capability to record user actions and customize them into a generic performance test script

- A sufficient capacity to dynamically adjust the user load introduced into the testing infrastructure

- Support for different networks protocols such as HTTPS, SOAP, SSH, and SFTP

- Tool stability at relatively high user loads, especially when trying to execute stress tests to determine the breaking point of the application

- The presence of support from the tool vendor and the community at large

- The existence of resources skilled in the tool to undertake the performance testing effort from scratch and get it to a closure

Next, let us look at similar tool selection criteria for API testing.

Selecting an API testing tool

Modern cloud platforms with distributed architecture comprising shared APIs are a common occurrence in every organization. There is a stronger push than ever to move all the business logic into the API layer, keeping the frontend light and elegant. This makes API testing crucial to deliver business value. There is a constant need to validate the accuracy of the API logic and fix any bugs as quickly as possible. With these factors in mind, let us go through certain criteria for API tool selection:

- The ability to set up and execute API calls quickly and easily
- The capability to create test suites and automatically run them
- Options to mock an API call
- The ability to create a chain of API calls to simulate a business workflow
- Options to customize the test suite by adding additional scripting wherever needed
- Support for non-functional testing
- Support for various technologies such as REST, SOAP, messaging systems (Kafka), and databases
- The presence of thorough documentation
- Capabilities in free versus licensed versions

Even though extensive API testing provides sizeable test coverage, there are parts of the application that cannot be tested using APIs. This is where a good web testing tool comes in handy. Let us review the items to look for in a solid web testing tool.

Selecting a web testing tool

A good web testing tool provides more diverse test coverage than just validating the API logic. It confers the ability to set up tests addressing the complete system behavior. Some of the most important features of a good web testing tool are as follows:

- Support for specific platforms and technologies used by the application being tested
- Stability when setting up long-running end-to-end tests
- Support for capture and playback tests for getting tests up and running in a brief time
- The ability to monitor and inspect backend parts of the application while running frontend tests
- Support for constant improvements and documentation

Thus, a good web testing tool must be well rounded in its coverage to address end-to-end aspects of the application. Next, let us look at the factors that influence a mobile testing tool selection.

Mobile

A major chunk of the factors that we have looked at so far apply to mobile testing tools as well. Here, in this section, let us list the specific criteria that apply to mobile testing:

- Support for cross-device and cross-platform test setup and execution
- Support for custom scripting to extend the existing framework
- Integrates seamlessly with cloud-based mobile device emulators/simulators
- The ability to debug and pinpoint errors on various mobile platforms

In the next section, let us examine the common factors that play a key role in a user settling on a particular tool.

Common considerations

Every kind of testing demands its own set of standards in a test automation tool. But there are certain common details that must be analyzed upfront in this process. Let us examine them in this section.

CI/CD integration

Irrespective of the tool or the type of testing being performed, integration with a CI/CD tool is an absolute must in the current software engineering landscape. The tool's ability to integrate with the current development infrastructure with ease is critical. Sometimes, the introduction of a new tool in the testing ecosystem demands software engineers to change the way they deliver software. The tool should offer plugins or other installable tools to blend with the CI/CD pipeline, thus paving the way for engineers to focus on writing the tests, with the tool taking care of the integration automatically. In fact, a testing tool should be an enabler of strong CI/CD processes if the team does not already have one.

Next, let us review some budget considerations while selecting a test automation tool.

Knowing your budget

Usually, one of the chief aspects of deciding on a test automation is its cost. Deciding between open source and building one from scratch is usually the first step. This depends immensely on the available time and resources at disposal. There should also be an upfront analysis done on the diverse options available in the licensed version to evaluate the fit. It is useful to analyze the risks involved in not adopting a tool and build on the benefits the tool will eventually offer. It is also important to assess how the licensing cost will change over time as the engineering organization evolves. It is vital to note that the introduction of a test automation tool would increase the costs initially, but eventually should reduce the quality and deployment costs when rightly integrated and used with the DevOps processes.

Let us next look at how having the resources with the right skills accelerates tool adoption.

Knowing the team's skillset

Even a well-rounded testing tool that is not utilized to its full potential will not deliver the promised ROI. Having the skilled resources to use the testing tool makes an enormous difference in the test automation effort. It is crucial to get the team's feedback about the tool selection process to make sure the tool addresses all the pain points faced by the team. Team members should be comfortable with the scripting language supported by the tool, and proper training must be provided to address the knowledge gaps. It also helps to select a language that is already being used by the software engineers on the team. This gives the team access to more coding experts. Quality engineers should be involved in the code review process, which would assist them in learning the intricacies and best practices of the programming language used. It would be wise to produce a tool onboarding plan and get it reviewed by all the users.

Summary

In this chapter, we looked at and understood some of the chief tools used for quality engineering on a daily basis. We started with the CLI and looked at some of the commonly used commands. Then, we explored what Git is and familiarized ourselves with basic Git commands to initialize and make changes to a code repository. We continued to review common test automation frameworks and their high-level architecture. We gained a solid understanding of the purpose of each of these tools/frameworks. In the concluding section, we reviewed the tool selection criteria for various kinds of testing and some common factors that influence testing tool selection.

Based on the contents of what we have seen in this chapter, you should be able to comfortably work on the CLI and should also feel ready to contribute via Git. You also gained a thorough understanding of the commonly used test automation frameworks and the analysis that goes before adopting one. In the next chapter, we will dive a little deeper into Git commands. We will also learn how to set up an IDE and familiarize ourselves with JavaScript.

Questions

1. What is the CLI and why is it important?
2. What is Git and how is it used?
3. What is Selenium and what are its components?
4. What is Appium mainly used for?
5. How is Cypress different from Selenium?
6. What is JMeter primarily used for?
7. How is AXE used in accessibility testing?
8. What are some common considerations when choosing a test automation tool?

Part 2:
Practical Affairs

We will begin this part by familiarizing ourselves with the basics of JavaScript and some advanced topics in Git. Subsequently, you will take up various test automation tools hands-on and learn how to quickly get up and running with them. You will familiarize yourself with their setup and how to use them to execute a basic test. By the end of this section, you will have acquired the skills to take up test automation for various platforms.

This part has the following chapters:

- *Chapter 4, Getting Started with the Basics*
- *Chapter 5, Test Automation for Web*
- *Chapter 6, Test Automation for Mobile*
- *Chapter 7, Test Automation for APIs*
- *Chapter 8, Test Automation for Performance*

Getting Started with the Basics

In the previous chapter, *Common Tools and Frameworks*, we gained sufficient context on the various tools that an automation engineer will use day-to-day. We will extend that journey in this chapter by diving deeper into some advanced Git topics in the first part. Then, we will look at how to download and set up an **Integrated Development Environment** (**IDE**). Finally, we will expand our learning to include the basics of JavaScript. The following is the list of topics we will be going over:

- Getting more familiar with Git

- Using an IDE

- Introduction to JavaScript

Technical requirements

In this chapter, we will continue looking at working examples of Git through the **Command-Line Interface** (**CLI**). We will be using the **Terminal** software on the Mac for our examples. Windows users can use the built-in **PowerShell** to execute these commands. We will also be downloading and exploring VS Code, which is an IDE. Please check this page for the download requirements: `https://code.visualstudio.com/docs/supporting/requirements`. We also expect you to know the basics of HTML to follow along with the next section on JavaScript.

All the code snippets can be found in the GitHub repository: `https://github.com/PacktPublishing/B19046_Test-Automation-Engineering-Handbook` in the `src/ch4` folder.

Getting more familiar with Git

For quality engineers, being proficient in Git and its usage is as important as their knowledge of test automation frameworks. Committing changes and managing branches are critical tasks that they should know by rote. It not only boosts their productivity but also makes the collaboration with the rest of the engineers seamless. In the next section, let us look a little deeper into committing a change and what goes into a commit message.

Committing a change

In the previous chapter, we briefly looked at what a Git `commit` command does. The `commit` command helps save all the changes made in the local repository.

Now, let's look at the importance of writing a descriptive commit message.

Importance of a commit message

More often than not, every software engineer views the commit history to understand what changes went in and why they were made. Anyone viewing a commit message should be able to get adequate context on why a certain change was made. If the commit messages are incomplete, then the engineers have to endure the tedious process of going through each code difference in a commit. For a large and complex commit, this can be very cumbersome. It is, therefore, tremendously important to get into the habit of taking the time to write clean and concise commit messages.

There are a few commonly used flags with this command. Let us look at them in action next.

Commonly used flags

We will look at three of the frequently used flags with the `commit` command in this section:

- `-m` This flag associates a message with the commit. We learned about the usage of this flag in the previous chapter. *Figure 4.1* illustrates adding multiline comments using the `git commit` command with the `-m` flag:

```
→ B19046_Test-Automation-Engineering-Handbook git:(main) git checkout -b ch4/git_commit_multiline
Switched to a new branch 'ch4/git_commit_multiline'
→ B19046_Test-Automation-Engineering-Handbook git:(ch4/git_commit_multiline) git status
On branch ch4/git_commit_multiline
Changes not staged for commit:
  (use "git add <file>..." to update what will be committed)
  (use "git restore <file>..." to discard changes in working directory)
        modified:   README.md

no changes added to commit (use "git add" and/or "git commit -a")
→ B19046_Test-Automation-Engineering-Handbook git:(ch4/git_commit_multiline) ✗ git add README.md
→ B19046_Test-Automation-Engineering-Handbook git:(ch4/git_commit_multiline) ✗ git commit -m "readme file: add chapter 4 header
" -m "readme file: add a note for git commit multiline comment"
[ch4/git_commit_multiline a858d24] readme file: add chapter 4 header
 1 file changed, 3 insertions(+), 1 deletion(-)
→ B19046_Test-Automation-Engineering-Handbook git:(ch4/git_commit_multiline) git push origin ch4/git_commit_multiline
Enumerating objects: 5, done.
Counting objects: 100% (5/5), done.
Delta compression using up to 8 threads
Compressing objects: 100% (3/3), done.
Writing objects: 100% (3/3), 437 bytes | 437.00 KiB/s, done.
Total 3 (delta 1), reused 0 (delta 0), pack-reused 0
remote: Resolving deltas: 100% (1/1), completed with 1 local object.
remote:
remote: Create a pull request for 'ch4/git_commit_multiline' on GitHub by visiting:
remote:      https://github.com/PacktPublishing/B19046_Test-Automation-Engineering-Handbook/pull/new/ch4/git_commit_multiline
remote:
To https://github.com/PacktPublishing/B19046_Test-Automation-Engineering-Handbook.git
 * [new branch]      ch4/git_commit_multiline -> ch4/git_commit_multiline
→ B19046_Test-Automation-Engineering-Handbook git:(ch4/git_commit_multiline) █
```

Figure 4.1 – git commit multiline comment

Figure 4.2 shows the view of this commit in GitHub:

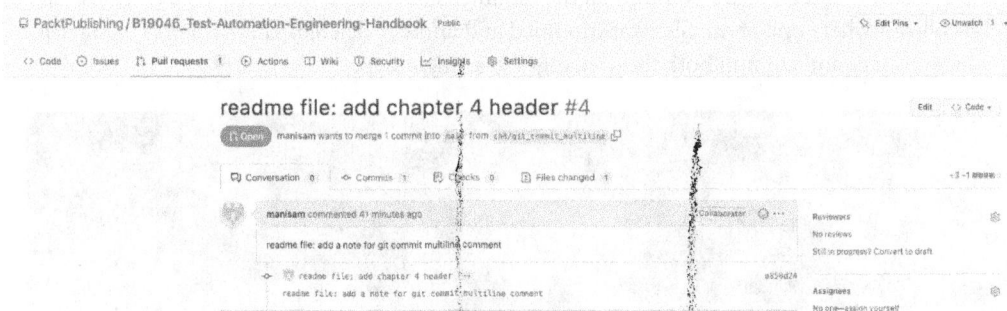

Figure 4.2 – GitHub multiline commit view

- --amend: This flag helps modify the most recent commit in the local repository instead of creating a new commit. *Figure 4.3* demonstrates an example where a new file was added in an initial commit. Then, using the --amend and --no-edit flags, a second file is added to complete the commit. The --no-edit flag allows you to complete the commit without editing the commit message:

```
→  B19046_Test-Automation-Engineering-Handbook git:(ch4/git_amend) x git status
On branch ch4/git_amend
Untracked files:
  (use "git add <file>..." to include in what will be committed)
        src/ch4/

nothing added to commit but untracked files present (use "git add" to track)
→  B19046_Test-Automation-Engineering-Handbook git:(ch4/git_amend) x git add src/ch4/
→  B19046_Test-Automation-Engineering-Handbook git:(ch4/git_amend) x git commit -m "Add ch4 folder"
[ch4/git_amend 03f2f9e] Add ch4 folder
 1 file changed, 0 insertions(+), 0 deletions(-)
 create mode 100644 src/ch4/git_amend_1.txt
→  B19046_Test-Automation-Engineering-Handbook git:(ch4/git_amend) git status
On branch ch4/git_amend
Untracked files:
  (use "git add <file>..." to include in what will be committed)
        src/ch4/git_amend_2.txt

nothing added to commit but untracked files present (use "git add" to track)
→  B19046_Test-Automation-Engineering-Handbook git:(ch4/git_amend) x git add src/ch4/git_amend_2.txt
→  B19046_Test-Automation-Engineering-Handbook git:(ch4/git_amend) x git commit --amend --no-edit
[ch4/git_amend 6d88963] Add ch4 folder
 Date: Sat Aug 20 21:54:38 2022 -0700
 2 files changed, 0 insertions(+), 0 deletions(-)
 create mode 100644 src/ch4/git_amend_1.txt
 create mode 100644 src/ch4/git_amend_2.txt
→  B19046_Test-Automation-Engineering-Handbook git:(ch4/git_amend)
```

Figure 4.3 – git commit with the --amend flag

- -a: This flag automatically commits all the modified and deleted files. *Figure 4.4* demonstrates a commit where one of the files was modified and another deleted. The –a flag is being used here to stage and commit both these changes in a single step:

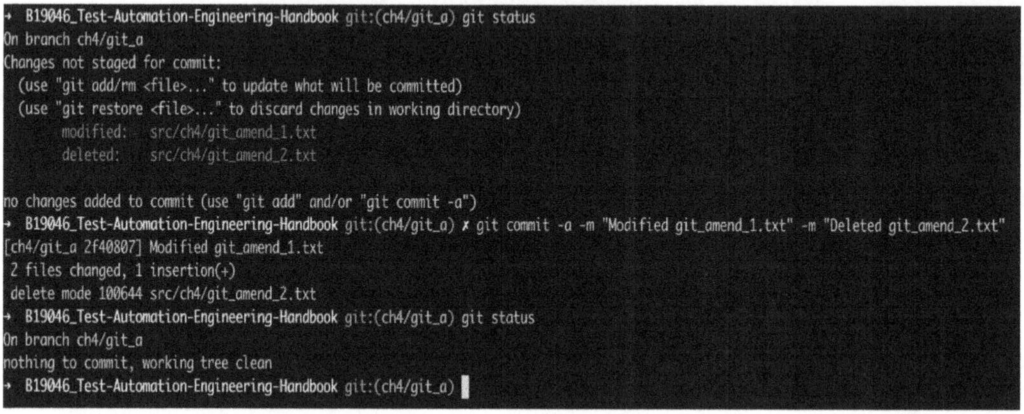

```
→  B19046_Test-Automation-Engineering-Handbook git:(ch4/git_a) git status
On branch ch4/git_a
Changes not staged for commit:
  (use "git add/rm <file>..." to update what will be committed)
  (use "git restore <file>..." to discard changes in working directory)
        modified:   src/ch4/git_amend_1.txt
        deleted:    src/ch4/git_amend_2.txt

no changes added to commit (use "git add" and/or "git commit -a")
→  B19046_Test-Automation-Engineering-Handbook git:(ch4/git_a) x git commit -a -m "Modified git_amend_1.txt" -m "Deleted git_amend_2.txt"
[ch4/git_a 2f40807] Modified git_amend_1.txt
 2 files changed, 1 insertion(+)
 delete mode 100644 src/ch4/git_amend_2.txt
→  B19046_Test-Automation-Engineering-Handbook git:(ch4/git_a) git status
On branch ch4/git_a
nothing to commit, working tree clean
→  B19046_Test-Automation-Engineering-Handbook git:(ch4/git_a)
```

Figure 4.4 – git commit with the -a flag

In the next section, let us look at an example of resolving a merge conflict when pushing and pulling changes to/from the remote repository.

Resolving merge conflicts

Merge conflicts happen when changes have transpired in the same region of a file and Git cannot automatically merge the changes in the file. It is possible that two different engineers are working on the same file and tried to push their changes to the remote repository. In such cases, Git fails the merge processes and forces manual resolution of the merge conflict. Without an IDE, this process can get really messy easily and might end up consuming a lot of the programmer's time. Resolving merge conflicts without an IDE usually involves viewing/editing multiple files through a CLI editor and identifying and fixing the parts of the file that are in conflict. This is a tedious process, but IDEs provide an interface to deal with conflicts and it is usually completed with a few clicks after manual file inspection.

Let us now look at how to resolve a merge conflict step by step. *Figure 4.5* illustrates a ch4/merge_conflict_branch_1 branch where the git_amend_2.txt file was updated and this change was pushed to the remote repository and merged with main through a pull request.

```
→ B19046_Test-Automation-Engineering-Handbook git:(ch4/merge_conflict_branch_1) git status
On branch ch4/merge_conflict_branch_1
Changes not staged for commit:
  (use "git add <file>..." to update what will be committed)
  (use "git restore <file>..." to discard changes in working directory)
        modified:   src/ch4/git_amend_2.txt

no changes added to commit (use "git add" and/or "git commit -a")
→ B19046_Test-Automation-Engineering-Handbook git:(ch4/merge_conflict_branch_1) ✗ git add src/ch4/git_amend_2.txt
→ B19046_Test-Automation-Engineering-Handbook git:(ch4/merge_conflict_branch_1) ✗ git commit -m "Update git_amend_2.txt"
[ch4/merge_conflict_branch_1 ed13825] Update git_amend_2.txt
 1 file changed, 1 insertion(+)
→ B19046_Test-Automation-Engineering-Handbook git:(ch4/merge_conflict_branch_1) git push origin ch4/merge_conflict_branch_1
Enumerating objects: 9, done.
Counting objects: 100% (9/9), done.
Delta compression using up to 8 threads
Compressing objects: 100% (4/4), done.
Writing objects: 100% (5/5), 445 bytes | 445.00 KiB/s, done.
Total 5 (delta 1), reused 0 (delta 0), pack-reused 0
remote: Resolving deltas: 100% (1/1), completed with 1 local object.
remote:
remote: Create a pull request for 'ch4/merge_conflict_branch_1' on GitHub by visiting:
remote:      https://github.com/PacktPublishing/B19046_Test-Automation-Engineering-Handbook/pull/new/ch4/merge_conflict_branch_1
remote:
To https://github.com/PacktPublishing/B19046_Test-Automation-Engineering-Handbook.git
```

Figure 4.5 – The git_amend_2.txt file updated and merged

In another branch, let us try to merge the changes from *Figure 4.4*, where the `git_amend_2.txt` file was deleted. It is evident that these two changes contradict each other. *Figure 4.6* shows the merged changes being fetched from the `main` branch:

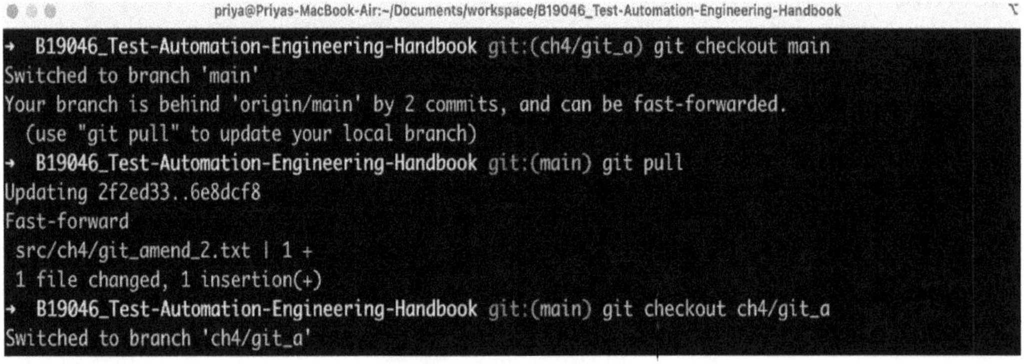

```
priya@Priyas-MacBook-Air:~/Documents/workspace/B19046_Test-Automation-Engineering-Handbook

→ B19046_Test-Automation-Engineering-Handbook git:(ch4/git_a) git checkout main
Switched to branch 'main'
Your branch is behind 'origin/main' by 2 commits, and can be fast-forwarded.
  (use "git pull" to update your local branch)
→ B19046_Test-Automation-Engineering-Handbook git:(main) git pull
Updating 2f2ed33..6e8dcf8
Fast-forward
 src/ch4/git_amend_2.txt | 1 +
 1 file changed, 1 insertion(+)
→ B19046_Test-Automation-Engineering-Handbook git:(main) git checkout ch4/git_a
Switched to branch 'ch4/git_a'
```

Figure 4.6 – Fetched merged changes from main

Figure 4.7 shows the result when the conflicting branch is being rebased with the main. Rebasing is the process of combining a chain of commits and applying it on top of a new base commit. Git automatically creates the new commit and applies it on the current base. Frequent rebasing from the main/master branch helps keep a sequential project history. At this point in the process, the conflicts have to be resolved manually. The engineer has to look through the file and accept or reject others' changes based on the project's needs.

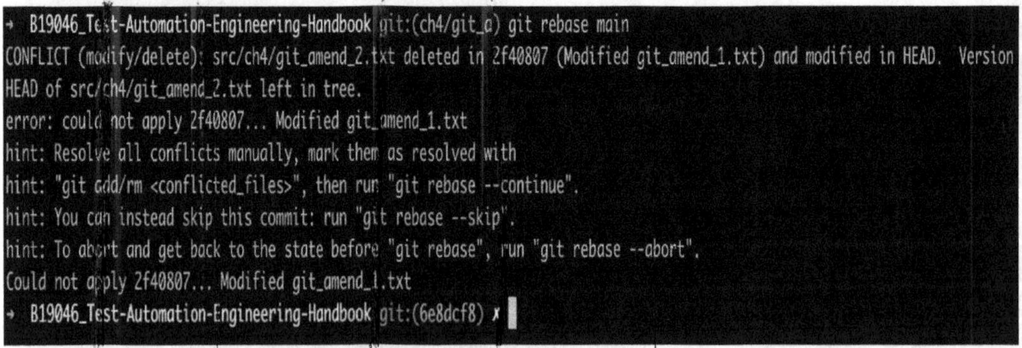

```
→ B19046_Test-Automation-Engineering-Handbook git:(ch4/git_a) git rebase main
CONFLICT (modify/delete): src/ch4/git_amend_2.txt deleted in 2f40807 (Modified git_amend_1.txt) and modified in HEAD.  Version
HEAD of src/ch4/git_amend_2.txt left in tree.
error: could not apply 2f40807... Modified git_amend_1.txt
hint: Resolve all conflicts manually, mark them as resolved with
hint: "git add/rm <conflicted_files>", then run "git rebase --continue".
hint: You can instead skip this commit: run "git rebase --skip".
hint: To abort and get back to the state before "git rebase", run "git rebase --abort".
Could not apply 2f40807... Modified git_amend_1.txt
→ B19046_Test-Automation-Engineering-Handbook git:(6e8dcf8) ✗
```

Figure 4.7 – Merge conflict message

In this case, let's resolve the merge conflict by accepting the incoming changes from the remote rather than pushing the delete. This is simpler when done through an IDE, as shown in *Figure 4.8*. On navigating to the source control pane in the IDE and by right clicking the file, the user is shown an option to accept the incoming changes:

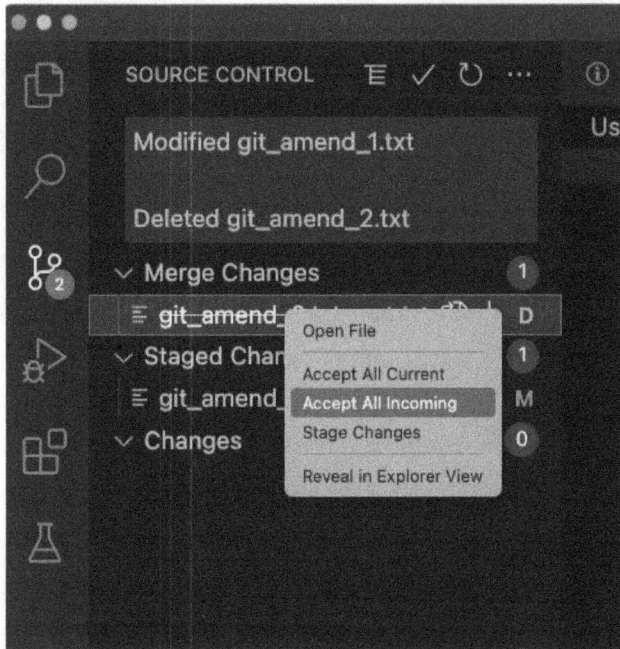

Figure 4.8 – Accept incoming changes

Figure 4.9 shows the result of staging the accepted changes, which results in the deleted file being retained with the modified contents from the main branch:

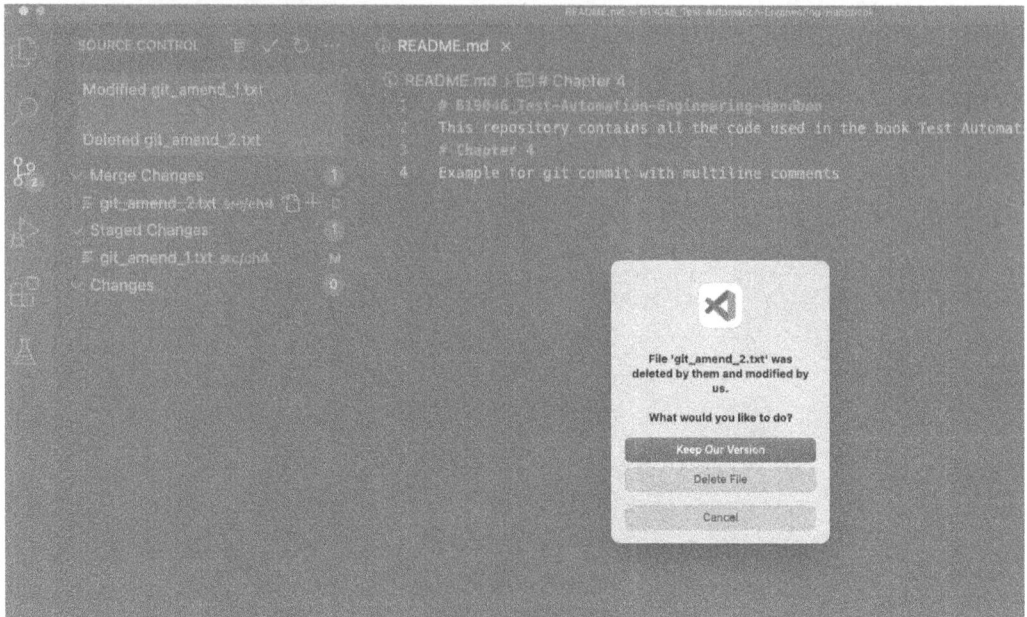

Figure 4.9 – Stage the accepted changes

Now that the merge conflict is resolved, the rebase can be continued using `git rebase --continue` to complete the commit and merge process subsequently. It is important to remember to pull from remote or other branches (if necessary) before beginning any new work on the local code base. This keeps the branch updated, thereby reducing merge conflicts. It is also vital to have continued communication with the rest of the team when deciding which changes to accept/reject when resolving merge conflicts. Next, let's look at a few additional Git commands that might help you save a lot of time.

Additional Git commands

Let us familiarize ourselves with a few additional commands that will help boost the productivity of the engineers:

- `git bisect`: This command helps identify a certain commit that is buggy or caused certain tests to fail. It uses binary search to narrow down the search to a single commit from hundreds of commits. The reference for the individual commits can be found using the `git log` command.

- `git cherry-pick`: This command helps pick changes from a specific branch and apply them to the local branch being worked on. It is very helpful in the case when you don't need to apply the changes that aren't relevant to your work or when you would like to test your local branch with changes from another remote branch.

- `git rebase`: This command was already used in some of the previous examples, and it helps simplify the merging process into the master or the main branch. It modifies the base of the local branch to the most recent commit in the main branch.

- `git revert`: This command provides a safe way to reverse the specified changes. Instead of deleting changes, it creates a new commit with inverse changes. This helps maintain revision history for future reference.

- `git reset`: This is a powerful command that helps move the repository back to an older commit by removing all the changes after that commit. It does this by moving the current head of the branch back to the older commit.

- `git reflog`: Git keeps track of changes to branches by maintaining reference logs, and this command helps to view those logs.

This concludes our section on the deeper dive into Git commands. Let us now start understanding what an IDE is and begin working with one.

Using an IDE

Quality engineering work demands working with various tools, and the right set of tools tends to make the engineer's job extremely efficient and productive. An IDE ranks among the top in helping engineers write and deploy high-quality code with speed and reliability. Before getting into the nitty-gritty of an IDE, let us try to understand what makes them so powerful. Apart from IDEs, some programmers prefer using text editors for writing code, and they are nothing more than tools that help edit text. Programmers regularly spend an enormous amount of time dealing with formatting and syntax details. There are text editors that come with syntax highlighting and color-coding that make their lives easier. These are usually apt for an experienced programmer. An IDE is a powerful tool that provides debugging, build management, and IntelliSensing features. It helps the programmer get up and running with a complex project really quickly. There is usually some learning curve with using an IDE and all its capabilities. But once we get familiarized with the features, it can be a great ally for our daily programming tasks. At the end of the day, it is the individual's preference on choosing to work with an editor or an IDE.

Let us consider, in the next section, some crucial factors in selecting an IDE.

Choosing an IDE

Not all IDEs are made equal, and there is a whole array of options available for engineers to choose from. Let us review some important features to look for in an IDE before settling on one:

- **GitHub integration**: This is a no-brainer and an absolute must-have since every engineer works with Git and needs integration with some form of remote code repository such as GitHub. This mainly provides the comfort of keeping the local code changes synchronized with the rest of the organization from within the IDE.

- **Debugging capabilities**: Running a program locally step-by-step to identify points of failure is a critical task that every engineer performs. IDEs should be able to support debugging in a variety of programming languages and provide options to view the runtime values of various aspects of the program, such as variables, objects, and so on.

- **Build tools integration**: The ability to package and execute all the dependencies from within the tool helps the quality engineers immensely in quickly making changes and retesting specific parts of the application. Integration with build tools reduces the split between the build process and testing the changes.

- **Plugins support**: There are numerous plugins usually available via a Marketplace to be downloaded and used within an IDE. These plugins or extensions provide valuable features such as intelligent code completion, live reload, bracket analyzer, and more, and they help save time on redundant tasks.

- **Cost versus speed**: A lot of the available IDEs are open source but sometimes need a compromise on the speed. It is important to gauge the scale of the project being worked on and choose a licensed or well-supported open source tool. The last thing an engineer would want is to work on an unstable IDE.

- **Coupling with languages/frameworks**: It is important to review which languages and technologies the IDE can support before selecting one. It should be compatible with the tech stack that the programmer will work on most of the time.

Let us now look at downloading and setting up VS Code, an open source IDE.

Downloading and setting up VS Code

In this section, we will be going over the process of downloading VS Code, which can run on macOS, Linux, and Windows. All the code examples cited in the rest of this book will use VS Code. You are free to use an IDE of your choice. Let us now go through the steps for manually installing VS Code on macOS. At the end of this section, you are provided with a shell script to perform this installation via the CLI:

1. Review this link for the necessary system requirements to download and set up VS Code: `https://code.visualstudio.com/docs/supporting/requirements`.

2. Click on the download link on the installation page to download the executable file:

   macOS and Mac: `https://code.visualstudio.com/docs/setup/mac`

   Linux: `https://code.visualstudio.com/docs/setup/linux`

   Windows: `https://code.visualstudio.com/docs/setup/windows`

3. Extract the downloaded archive file and move the **Visual Studio Code** application file to the `Applications` folder.

4. Double-click the **VS Code** icon in the `Applications` folder to open the VS Code IDE.

5. Use the **Settings** option in the **Preferences** menu for additional configuration. VS Code has inherent support for Git provided Git is installed on the machine.

   VS Code comes as a lightweight installation and in most cases, engineers would need additional components installed through the **Extensions** option in the **Preferences** menu.

6. Click the **Extensions** option and search for **Prettier,** which is a particularly useful tool for code formatting.

7. Select the **Prettier - Code formatter** extension and click on the **Install** button, as shown in *Figure 4.10*. This should complete the installation of the extension from the Marketplace.

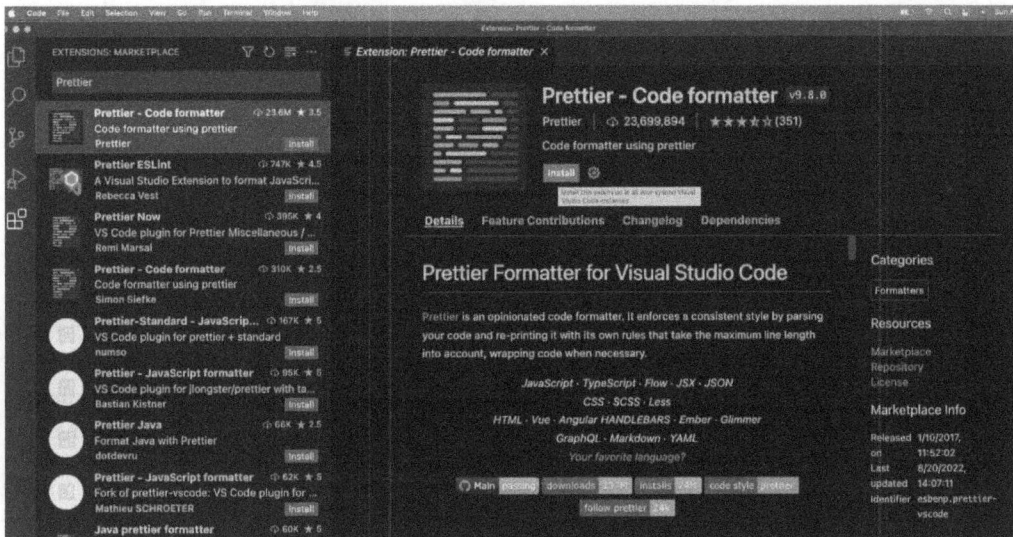

Figure 4.10 – Installing Prettier - Code formatter

There are hundreds of helpful extensions available for installation from the Marketplace. These extensions have a wide community of users supporting them, thereby creating a strong ecosystem. Users are strongly encouraged to browse through the available extensions and install them as necessary.

If you prefer using the CLI for installation, the following shell script can be used for installing VS Code on macOS. This example uses Homebrew (`https://brew.sh/`) for CLI installation:

```
#!/bin/sh


/bin/bash -c "$(curl -fsSL https://raw.githubusercontent.com/
Homebrew/install/HEAD/install.sh)"
```

```
brew tap homebrew/cask

brew install --cask visual-studio-code
```

This concludes our section on setting up the IDE. Let us now move on to learning the basics of JavaScript.

Introduction to JavaScript

JavaScript is the most common web programming language, and it is imperative for quality engineers to familiarize themselves with the basics. With minor exceptions, we will be using JavaScript to automate various levels of the application stack in the rest of this book. Let us start with the question of why we should learn JavaScript.

Why learn JavaScript?

One of the primary reasons to learn JavaScript is its complete integration with HTML/CSS and that it is enabled to work on all browsers. This makes it extremely easy to create and execute scripts. It is also a loosely typed language that provides engineers with great flexibility when writing applications. Its event-driven architecture helps engineers perform rigorous read/write operations with ease at a scale. Even though it was initially created to run only on browsers, it is currently being used extensively for building server-side applications. JavaScript, as a programming language, caters to beginner, intermediate, and advanced engineers. We will start with the basics in this section and learn about quite a few intermediate and advanced JavaScript topics throughout this book in the subsequent chapters. JavaScript also supports both functional and object-oriented programming paradigms. It is also useful to know that JavaScript conforms to **ECMAScript**, which is a standard used to ensure the interoperability of web pages between browsers.

Knowing all these advantages, let us start working on JavaScript in the next section.

Running a JavaScript program

In this section, let us explore how to run a basic JavaScript program from within our IDE. We will start by installing the **Node.js** runtime environment.

Installing Node.js

One of the easiest ways to execute a JavaScript program is to run it using Node.js. Node.js is a JavaScript runtime environment created to execute JavaScript code outside of a browser. Use `https://nodejs.org/en/` to download Node.js and navigate through the wizard to complete the installation of the latest stable version of Node.js. Check the installation using the `node --version` command.

Alternatively, you can use **Homebrew** on macOS or **winget** on Windows to download it via the CLI. At the time of writing this book, the latest stable node version is *18.12.1*:

```
→   cli_demo node --version
v18.12.1
→   cli_demo
```

Now that we have Node.js installed, let us execute our first JavaScript program.

Executing the JavaScript program

We will be using the VS Code IDE to execute the programs for the rest of this book. Let us go through step by step to execute a simple Hello World program written in JavaScript:

1. The console.log command outputs the message in the parenthesis to the console. Create a new hello_world.js file and save it with the contents, console.log('hello world');.

2. Open a new Terminal window by selecting the **New terminal** option from the **Terminal** menu. In the Terminal window, navigate to the folder in which the hello_world.js file exists and run the node hello_world.js command. This prints out the text hello world in the console:

```
→   B19046_Test-Automation-Engineering-Handbook git:(main)
X cd src/ch4
→   ch4 git:(main) X node hello_world.js
hello world
→   ch4 git:(main) X
```

We have set up our IDE and are able to execute a simple JavaScript program using Node.js. It is now time to get started with the basics.

JavaScript basics

Let us start by adding comments in a JavaScript file. Single-line comments can be added using // and multiline can be added by enclosing within /* */, as shown in *Figure 4.11*:

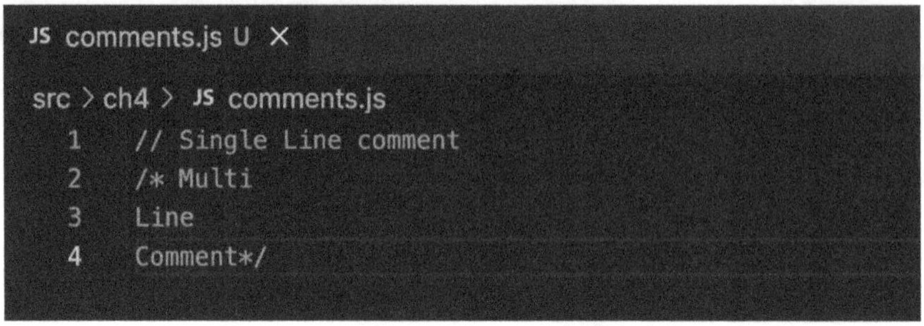

```
JS comments.js U ✕

src > ch4 > JS comments.js
    1    // Single Line comment
    2    /* Multi
    3    Line
    4    Comment*/
```

Figure 4.11 – Adding comments in a JavaScript file

Let us look at variables in the next section.

Variables

Variables are references to the storage location. `var` was the standard declaration command until ES6. But in ES6, some of the scoping and hoisting issues with `var` were addressed, and the common ways to set variables in JavaScript became `let` and `const`. The main difference is that `let` allows the variable to be assigned a different value later in the program, but `const` does not. *Figure 4.12* demonstrates the code where the variable temperature is being created and reassigned using `let`. It also shows the error when attempting to reassign the value of `city`, which is declared as `const`:

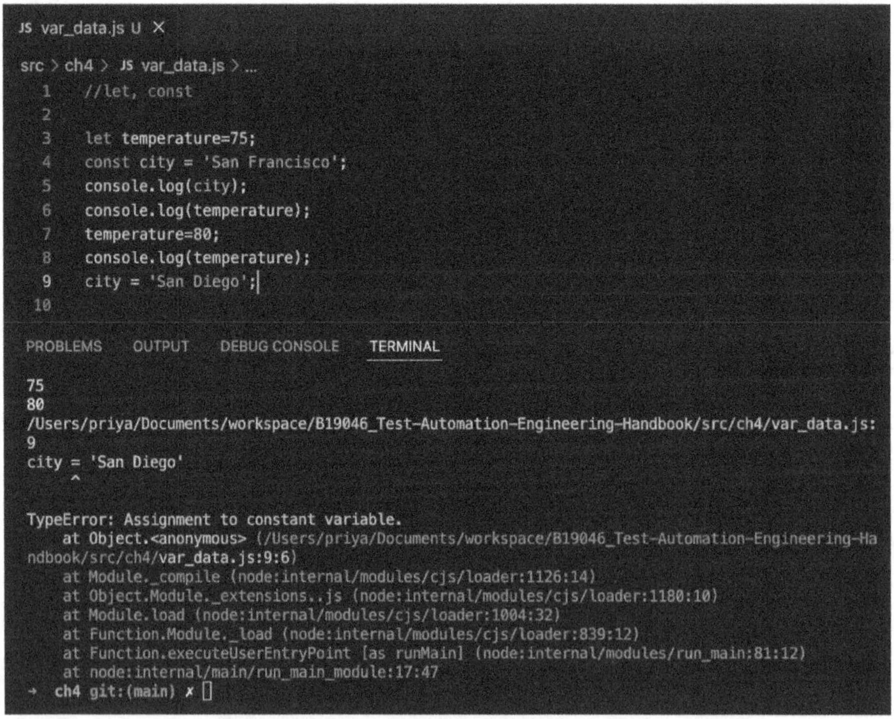

```
JS var_data.js U ✕

src > ch4 > JS var_data.js > ...
    1    //let, const
    2
    3    let temperature=75;
    4    const city = 'San Francisco';
    5    console.log(city);
    6    console.log(temperature);
    7    temperature=80;
    8    console.log(temperature);
    9    city = 'San Diego';|
   10

PROBLEMS    OUTPUT    DEBUG CONSOLE    TERMINAL

75
80
/Users/priya/Documents/workspace/B19046_Test-Automation-Engineering-Handbook/src/ch4/var_data.js:
9
city = 'San Diego'
       ^

TypeError: Assignment to constant variable.
    at Object.<anonymous> (/Users/priya/Documents/workspace/B19046_Test-Automation-Engineering-Ha
ndbook/src/ch4/var_data.js:9:6)
    at Module._compile (node:internal/modules/cjs/loader:1126:14)
    at Object.Module._extensions..js (node:internal/modules/cjs/loader:1180:10)
    at Module.load (node:internal/modules/cjs/loader:1004:32)
    at Function.Module._load (node:internal/modules/cjs/loader:839:12)
    at Function.executeUserEntryPoint [as runMain] (node:internal/modules/run_main:81:12)
    at node:internal/main/run_main_module:17:47
→ ch4 git:(main) ✗ []
```

Figure 4.12 – Variable types, let and const

> **Note**
>
> It is better to use `const` to declare variables unless there is a need to reassign the value of the variable.

Next, let us take a look at data types.

Data types

Data types fall into two categories: primitives and objects. In primitive data types, the data is assigned directly to memory. Let us look at the primitive type in this section and devote the next section to objects.

Figure 4.13 shows the code snippet with variable assignment for each of the primitive data types. `string`, `number`, `boolean`, `null`, and `undefined` form the most used primitive data types. Strings must be defined using double or single quotes. Boolean can contain only `true` or `false` values. Also, note that there isn't a separate data type for decimal values; it is also a number. Null means an empty variable with no value. Any variable that is initialized without a value has an `undefined` type. Variable types can be tested using `typeof` followed by the name of the variable:

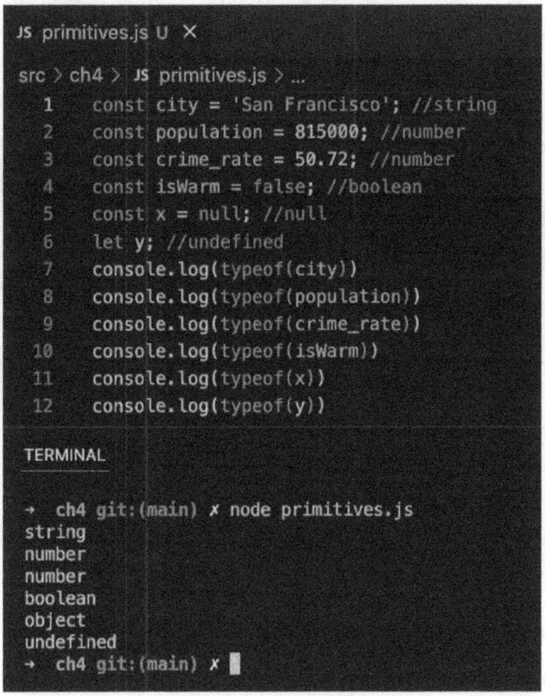

```
JS primitives.js U ✕

src > ch4 > JS primitives.js > ...
    1    const city = 'San Francisco'; //string
    2    const population = 815000; //number
    3    const crime_rate = 50.72; //number
    4    const isWarm = false; //boolean
    5    const x = null; //null
    6    let y; //undefined
    7    console.log(typeof(city))
    8    console.log(typeof(population))
    9    console.log(typeof(crime_rate))
   10    console.log(typeof(isWarm))
   11    console.log(typeof(x))
   12    console.log(typeof(y))

TERMINAL

→  ch4 git:(main) ✗ node primitives.js
string
number
number
boolean
object
undefined
→  ch4 git:(main) ✗ █
```

Figure 4.13 – Data types

> **Note**
>
> You might have noticed from *Figure 4.13* that `null` has a `typeof` object. This is considered a bug in JavaScript. The explanation for this can be found in this link: `https://developer.mozilla.org/en-US/docs/Web/JavaScript/Reference/Operators/typeof#typeof_null`.

In the next section, let us look at strings in a little more detail, as they are a widely used primitive data type.

Working with strings

There are a few common string usages and methods that engineers should be aware of.

Figure 4.14 illustrates the concatenation of two strings to produce a meaningful statement. Backticks can be used to enclose the string with a dollar sign followed by curly brackets with the variable name:

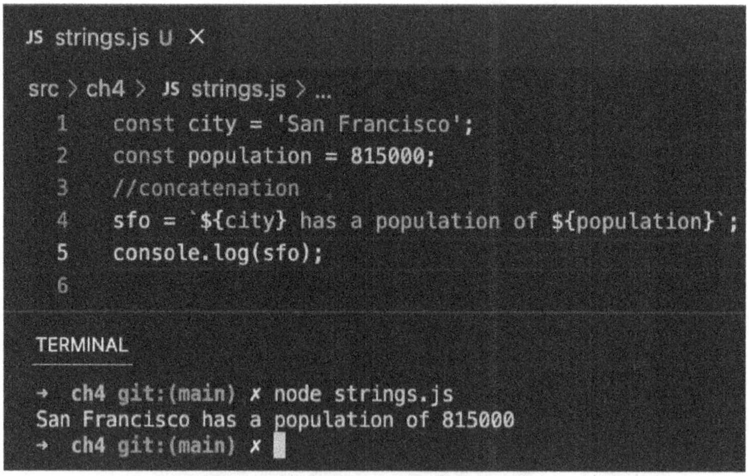

```
JS strings.js U ✕

src > ch4 > JS strings.js > ...
  1    const city = 'San Francisco';
  2    const population = 815000;
  3    //concatenation
  4    sfo = `${city} has a population of ${population}`;
  5    console.log(sfo);
  6

TERMINAL

→  ch4 git:(main) ✗ node strings.js
San Francisco has a population of 815000
→  ch4 git:(main) ✗ ▮
```

Figure 4.14 – String concatenation

Let us now familiarize ourselves more with strings by working on some properties and methods. A method is a function that is associated with an object. Methods must be invoked with parenthesis, while properties do not need them:

- `length`: This property returns the number of characters in the string
- `toUpperCase()`/`toLowerCase()`: These methods convert the given string to uppercase or lowercase
- `substring()`: This method takes in start and end indices and returns a substring matching the indices
- `split()`: This method splits the given string into arrays based on the provided splitter parameter

Figure 4.15 summarizes how these properties and methods are used with strings:

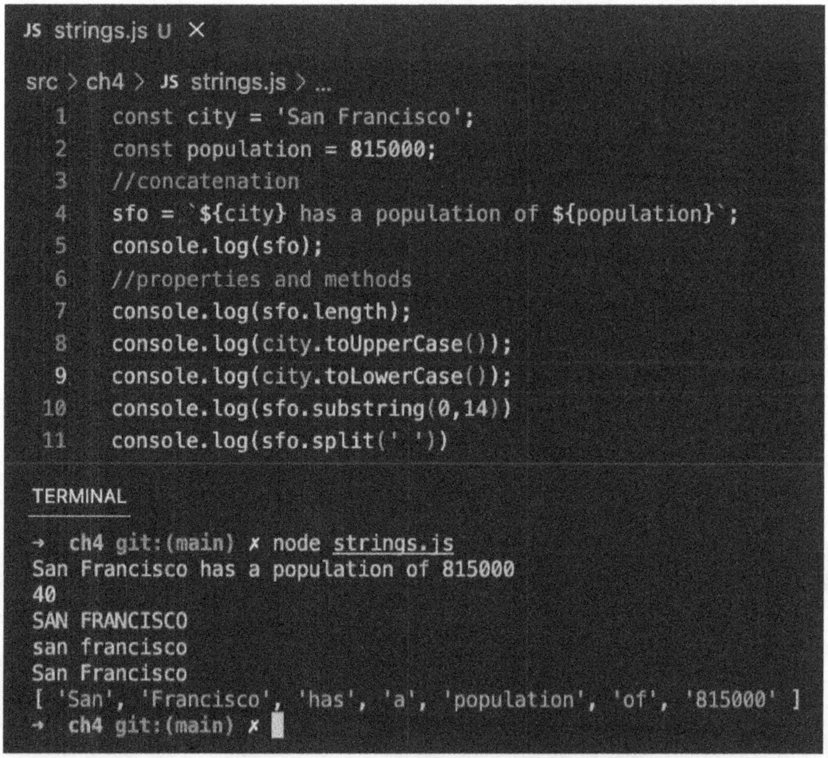

```
JS strings.js U ✕

src > ch4 > JS strings.js > ...
    1    const city = 'San Francisco';
    2    const population = 815000;
    3    //concatenation
    4    sfo = `${city} has a population of ${population}`;
    5    console.log(sfo);
    6    //properties and methods
    7    console.log(sfo.length);
    8    console.log(city.toUpperCase());
    9    console.log(city.toLowerCase());
   10    console.log(sfo.substring(0,14))
   11    console.log(sfo.split(' '))

TERMINAL

→  ch4 git:(main) ✗ node strings.js
San Francisco has a population of 815000
40
SAN FRANCISCO
san francisco
San Francisco
[ 'San', 'Francisco', 'has', 'a', 'population', 'of', '815000' ]
→  ch4 git:(main) ✗ ▊
```

Figure 4.15 – Working with strings

With this introduction to variables and data types, let us now take on the topic of objects in JavaScript.

Getting to know the JavaScript objects

In JavaScript, an object is a collection of properties. Objects are first initialized using a variable name and assigned to a set of properties. Objects further provide methods to update these properties. Properties can be of any data type, sometimes even other objects. This enables building complex objects in JavaScript. Let us start learning about objects with arrays in the next section.

Using JavaScript arrays

Arrays are one of the most frequent data structures and are built-in objects in JavaScript. Arrays are nothing but variables that can hold multiple values. The first element of an array is indexed by 0 and the subsequent indices are incremented by 1. The size of the arrays can be changed by adding or deleting elements, and they can contain a mix of data types. Arrays can be initialized by enclosing the elements in square brackets, [ ]. Subsequently, the elements can be accessed by plugging the index within [ ]. Let's look at some commonly used array methods:

- `push(element)`: Adds an element at the end of the array.

- `unshift(element)`: Adds an element at the beginning of the array.

- `pop()`: Removes the last element of the array.

- `indexOf(element)`: Returns the index of the element in the array. Returns −1 if the element is not found in the array.

- `length()`: Returns the number of elements in the array.

Figure 4.16 shows these array operations in action:

```
JS arrays.js U  ✕

src > ch4 > JS arrays.js > ...
  1    //Arrays
  2    const cities = ['San Francisco', 'Los Angeles', 'San Diego', 'Irvine'];
  3    console.log(cities);
  4    console.log(cities[1]);
  5    cities.push('Oakland');
  6    console.log(cities);
  7    cities.unshift('Sunnyvale');
  8    console.log(cities);
  9    cities.push(1);
 10    console.log(cities);
 11    cities.pop();
 12    console.log(cities);
 13    console.log(cities.indexOf('San Diego'));
 14    console.log(cities.indexOf(1));
 15    console.log(cities.length);
```

Figure 4.16 – Array operations

Figure 4.17 shows the corresponding outputs. We begin by creating the array and printing it to the console. Then, we add elements to the end and beginning of the array. Subsequently, we work with the indices of the array, and finally, get the length of the array:

```
TERMINAL

→  ch4 git:(main) x node arrays.js
[ 'San Francisco', 'Los Angeles', 'San Diego', 'Irvine' ]
Los Angeles
[ 'San Francisco', 'Los Angeles', 'San Diego', 'Irvine', 'Oakland' ]
[
  'Sunnyvale',
  'San Francisco',
  'Los Angeles',
  'San Diego',
  'Irvine',
  'Oakland'
]
[
  'Sunnyvale',
  'San Francisco',
  'Los Angeles',
  'San Diego',
  'Irvine',
  'Oakland',
  1
]
[
  'Sunnyvale',
  'San Francisco',
  'Los Angeles',
  'San Diego',
  'Irvine',
  'Oakland'
]
3
-1
6
```

Figure 4.17 – Array operation outputs

Unlike a lot of other programming languages, arrays in JavaScript do not throw an `Array Out of Bounds` error when trying to access an index greater than or equal to the length of the array. JavaScript simply returns `undefined` when trying to access the non-existent index array. Arrays come with a wide variety of built-in methods, and I would strongly encourage you to browse through them at `https://developer.mozilla.org/en-US/docs/Web/JavaScript/Reference/Global_Objects/Array`. Next, let us look at work with object literals.

Working with object literals

Object literals allow properties to be defined as key-value pairs. Values of the properties can be other objects as well. Dot (.) or square bracket ([]) notations can be used to retrieve the value of a property. The code snippet demonstrated in *Figure 4.18* shows an object and array nested within the `movie` object. In such cases, the object name can be chained subsequently with the call to a nested data structure. Adding an extra property to the object is very simple and looks like a variable assignment. To further our example, it would be `movie['producer']='Danny DeVito`:

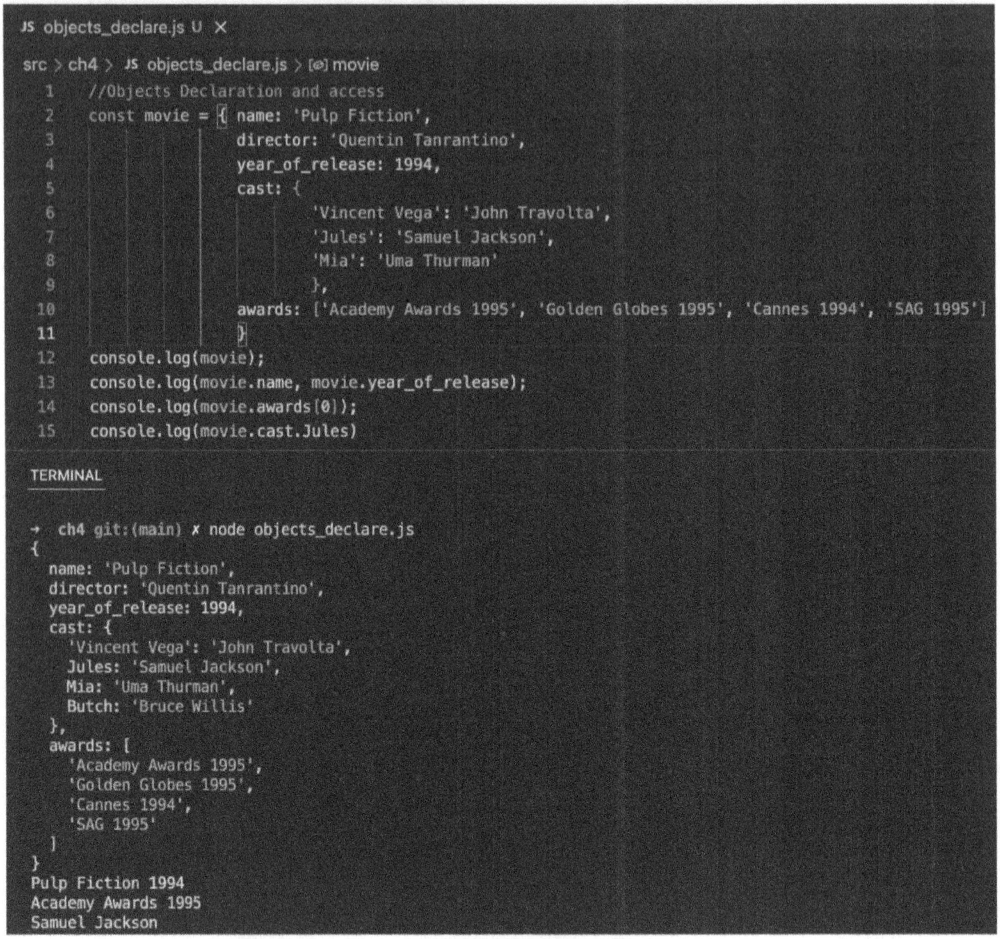

```
JS objects_declare.js ∪ ✕

src > ch4 > JS objects_declare.js > [∅] movie
   1    //Objects Declaration and access
   2    const movie = { name: 'Pulp Fiction',
   3                    director: 'Quentin Tanrantino',
   4                    year_of_release: 1994,
   5                    cast: {
   6                            'Vincent Vega': 'John Travolta',
   7                            'Jules': 'Samuel Jackson',
   8                            'Mia': 'Uma Thurman'
   9                            },
  10                    awards: ['Academy Awards 1995', 'Golden Globes 1995', 'Cannes 1994', 'SAG 1995']
  11                    }
  12    console.log(movie);
  13    console.log(movie.name, movie.year_of_release);
  14    console.log(movie.awards[0]);
  15    console.log(movie.cast.Jules)

TERMINAL

→  ch4 git:(main) ✗ node objects_declare.js
{
  name: 'Pulp Fiction',
  director: 'Quentin Tanrantino',
  year_of_release: 1994,
  cast: {
    'Vincent Vega': 'John Travolta',
    Jules: 'Samuel Jackson',
    Mia: 'Uma Thurman',
    Butch: 'Bruce Willis'
  },
  awards: [
    'Academy Awards 1995',
    'Golden Globes 1995',
    'Cannes 1994',
    'SAG 1995'
  ]
}
Pulp Fiction 1994
Academy Awards 1995
Samuel Jackson
```

Figure 4.18 – Objects

`const` prevents reassigning the variable but does not prevent modifying values within an object:

```
const a = { message: "hello" };
a.message = "world"; // this will work
```

Having learned the basics of JavaScript objects, let us now look at how to destructure one.

Destructuring an object

Object destructuring is used in JavaScript to extract property values and assign them to other variables. It has various advantages such as assigning multiple variables in a single statement, accessing properties from nested objects, and assigning a default value when a property doesn't exist. We use the same example as in the previous section but as shown in *Figure 4.19*, we destructure the `movie` object by specifying the `name` and `awards` variables within { }. On the right-hand side of the expression, we specify the `movie` object. We could also assign them to a variable within the curly brackets to fetch data from nested objects. Object destructuring was introduced in ECMAScript 6 and prior to this, extracting and assigning properties in such a way required a lot of boilerplate code.

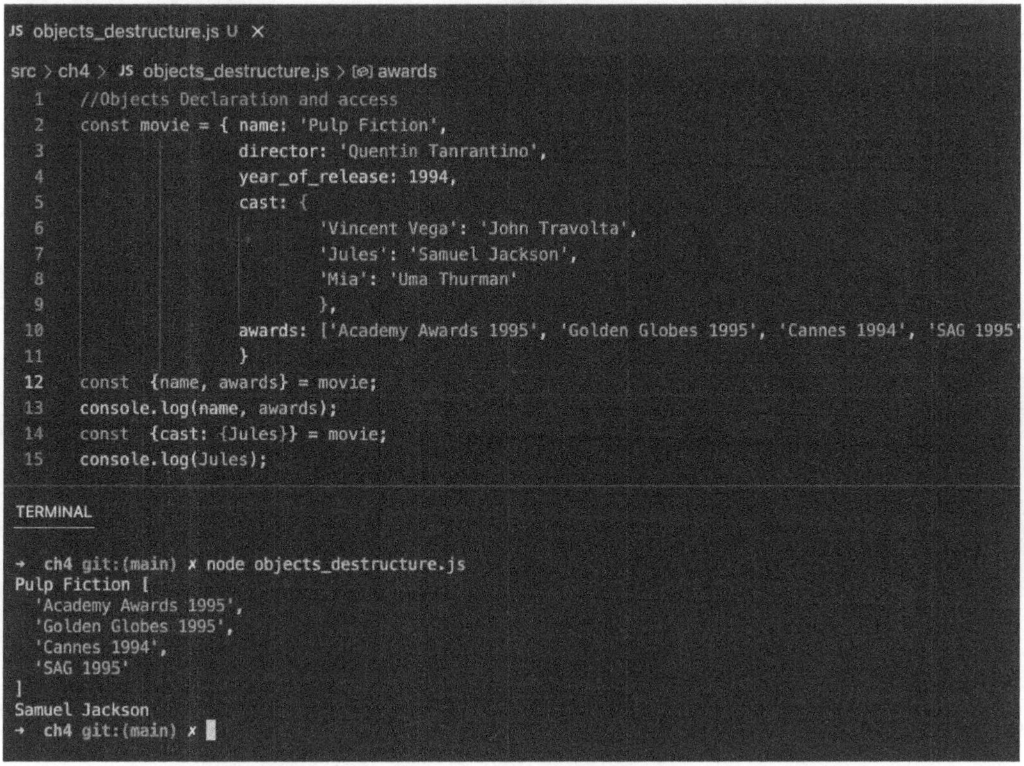

```
JS objects_destructure.js U ✕

src > ch4 > JS objects_destructure.js > [∅] awards
    1    //Objects Declaration and access
    2    const movie = { name: 'Pulp Fiction',
    3                    director: 'Quentin Tanrantino',
    4                    year_of_release: 1994,
    5                    cast: {
    6                            'Vincent Vega': 'John Travolta',
    7                            'Jules': 'Samuel Jackson',
    8                            'Mia': 'Uma Thurman'
    9                            },
   10                    awards: ['Academy Awards 1995', 'Golden Globes 1995', 'Cannes 1994', 'SAG 1995'
   11                    }
   12    const {name, awards} = movie;
   13    console.log(name, awards);
   14    const {cast: {Jules}} = movie;
   15    console.log(Jules);

TERMINAL

→  ch4 git:(main) ✗ node objects_destructure.js
Pulp Fiction [
  'Academy Awards 1995',
  'Golden Globes 1995',
  'Cannes 1994',
  'SAG 1995'
]
Samuel Jackson
→  ch4 git:(main) ✗ ▋
```

Figure 4.19 – Object destructuring

Let us next work with an array of objects.

Arrays of objects

Working with arrays of objects is crucial for quality engineers as a lot of times, the API response payloads in JSON format have multiple objects embedded within an array and their format is identical to JavaScript object literals. Let us consider an example where multiple `movie` objects are embedded within the `movies` array. We could use the `JSON.stringify` method to create a JSON string. The code snippet in *Figure 4.20* demonstrates how to access nested elements in an array of objects and how to create a JSON string from a JavaScript object:

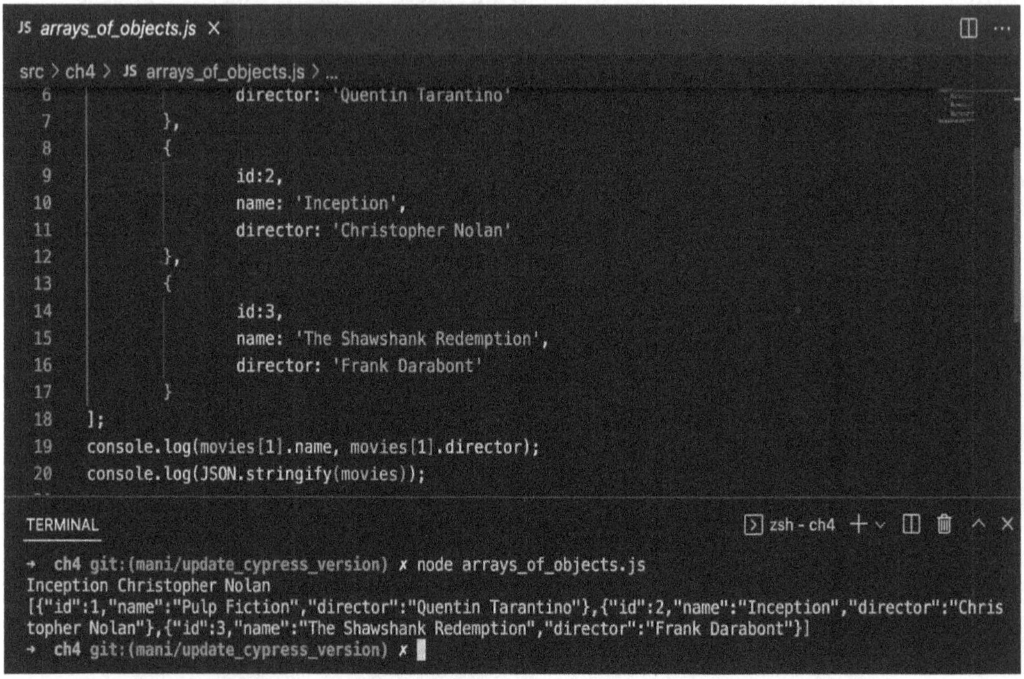

```
JS arrays_of_objects.js ×

src > ch4 > JS arrays_of_objects.js > ...
6                    director: 'Quentin Tarantino'
7            },
8            {
9                    id:2,
10                   name: 'Inception',
11                   director: 'Christopher Nolan'
12           },
13           {
14                   id:3,
15                   name: 'The Shawshank Redemption',
16                   director: 'Frank Darabont'
17           }
18   ];
19   console.log(movies[1].name, movies[1].director);
20   console.log(JSON.stringify(movies));

TERMINAL                                                zsh - ch4

→ ch4 git:(mani/update_cypress_version) x node arrays_of_objects.js
Inception Christopher Nolan
[{"id":1,"name":"Pulp Fiction","director":"Quentin Tarantino"},{"id":2,"name":"Inception","director":"Chris
topher Nolan"},{"id":3,"name":"The Shawshank Redemption","director":"Frank Darabont"}]
→ ch4 git:(mani/update_cypress_version) x
```

Figure 4.20 – Arrays of objects

In the next section, let us learn how to operate with loops and conditional statements.

Loops and conditionals

Loops and conditional statements form a basic pillar of any programming language. They help reduce runtime and make the program look cleaner. Let us look at each one of them and understand how they operate in the next sections, starting with loops.

Working with loops

One of the most frequently used loops is the `for` loop. A `for` loop contains three parameters: iterator assignment, condition, and increment. The code enclosed within the loop executes until the specified condition is met. In the simple example illustrated in *Figure 4.21*, we start with 0 and print the value of i until it meets the i<10 condition:

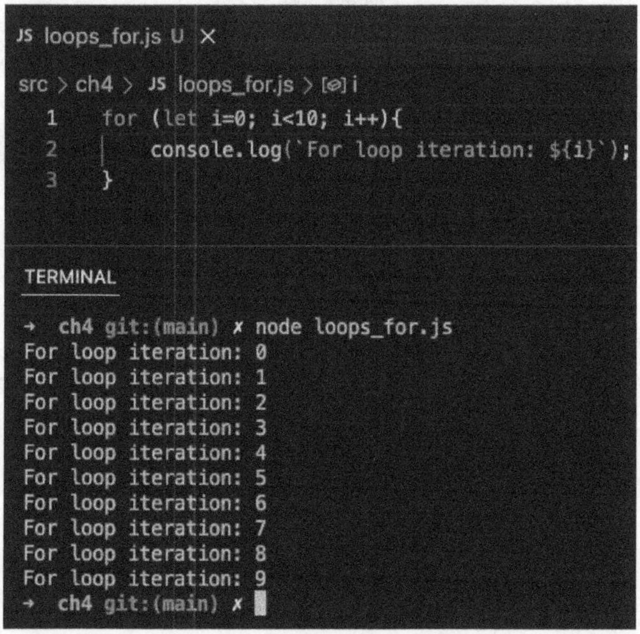

```
JS loops_for.js U ✕

src > ch4 > JS loops_for.js > [∅] i
   1    for (let i=0; i<10; i++){
   2        console.log(`For loop iteration: ${i}`);
   3    }



TERMINAL

→  ch4 git:(main) ✗ node loops_for.js
For loop iteration: 0
For loop iteration: 1
For loop iteration: 2
For loop iteration: 3
For loop iteration: 4
For loop iteration: 5
For loop iteration: 6
For loop iteration: 7
For loop iteration: 8
For loop iteration: 9
→  ch4 git:(main) ✗ ▊
```

Figure 4.21 – A simple for loop

> **Note**
>
> One of the common pitfalls while looping through an array is to accidentally exceed the last index of the array. The condition to check the array should be i<`array.length` or i<=`array.length-1`.

The `while` loop operates similarly to the `for` loop, but we set the variable outside of the loop. It is a common mistake to miss the increment or incorrectly specify the condition. Doing so would result in an infinite loop. *Figure 4.22* shows the same logic in the `while` loop:

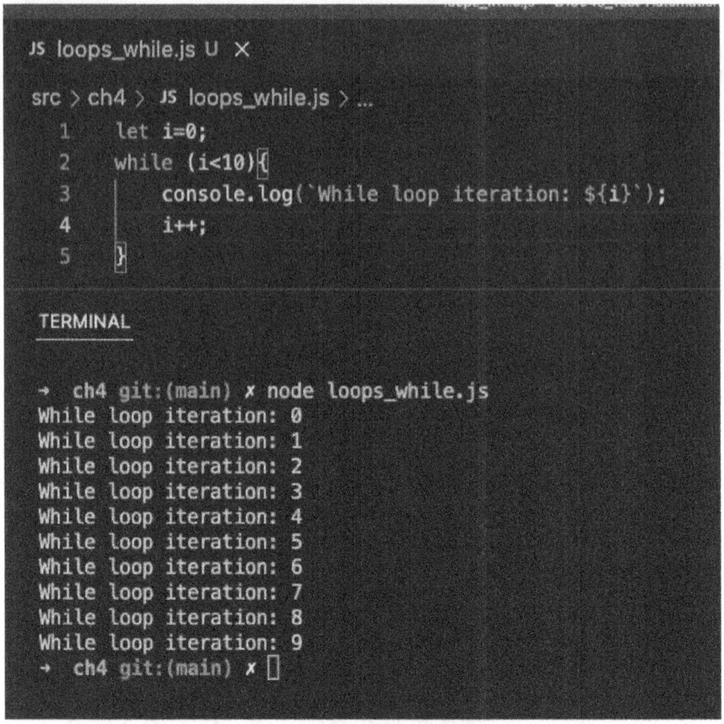

```
JS loops_while.js U ×

src > ch4 > JS loops_while.js > ...
    1    let i=0;
    2    while (i<10){
    3        console.log(`While loop iteration: ${i}`);
    4        i++;
    5    }


TERMINAL


→  ch4 git:(main) x node loops_while.js
While loop iteration: 0
While loop iteration: 1
While loop iteration: 2
While loop iteration: 3
While loop iteration: 4
While loop iteration: 5
While loop iteration: 6
While loop iteration: 7
While loop iteration: 8
While loop iteration: 9
→  ch4 git:(main) x □
```

Figure 4.22 – A simple while loop

Now, let us loop through the array of objects we created in *Figure 4.20*. For this purpose, we will use the for..of loop, which is much more readable than the regular for loop. In *Figure 4.23*, we have the code snippet that iterates over each of the objects in the movies array and prints the name and director. We create a temporary variable to hold the current entry of the array in the loop and use that variable to print the properties:

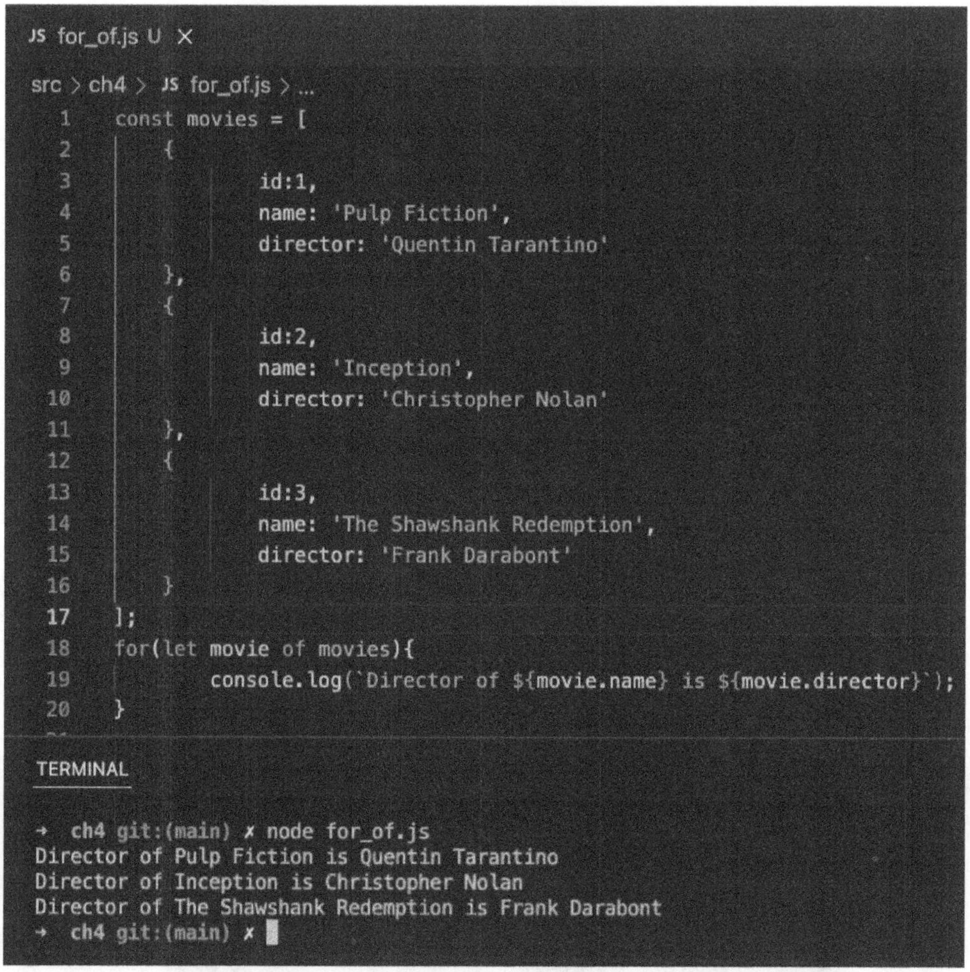

```js
JS for_of.js U X

src > ch4 > JS for_of.js > ...
  1    const movies = [
  2        {
  3                id:1,
  4                name: 'Pulp Fiction',
  5                director: 'Quentin Tarantino'
  6        },
  7        {
  8                id:2,
  9                name: 'Inception',
 10                director: 'Christopher Nolan'
 11        },
 12        {
 13                id:3,
 14                name: 'The Shawshank Redemption',
 15                director: 'Frank Darabont'
 16        }
 17    ];
 18    for(let movie of movies){
 19            console.log(`Director of ${movie.name} is ${movie.director}`);
 20    }
```

```
TERMINAL

→  ch4 git:(main) x node for_of.js
Director of Pulp Fiction is Quentin Tarantino
Director of Inception is Christopher Nolan
Director of The Shawshank Redemption is Frank Darabont
→  ch4 git:(main) x ▌
```

Figure 4.23 – A for..of loop

We looked at some very useful examples of loops in this section. Let us move on next to conditional statements.

Using the conditional statements

Conditional statements are used to separate the logic into different code blocks based on one or more conditions. The most common conditional statement is the if...else statement. This is better understood by referring to the code snippet in *Figure 4.24*. Here, we use conditional statements to assign a grade based on the student's score. We start with the if statement and check for the highest grade and then use a series of else if statements followed by the else statement to check for any score less than 60. The else if statements are useful to extend the logic to include additional conditions. It is important to remember that in the absence of an `else` statement, JavaScript ignores the conditional code block when the `if` condition is not `true`:

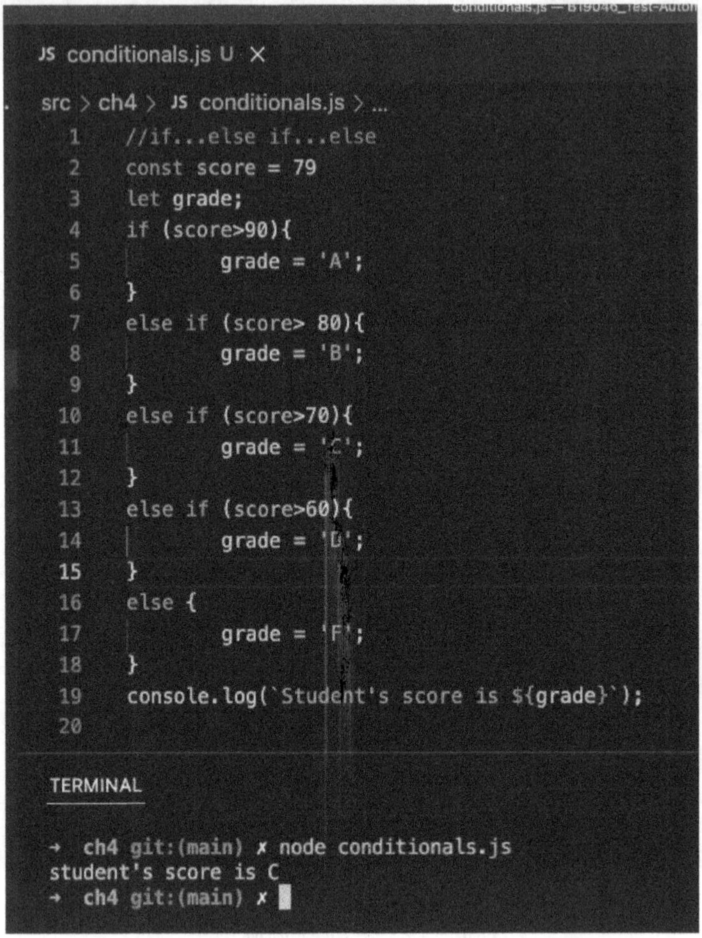
JS conditionals.js U X
src > ch4 > JS conditionals.js > ...
//if...else if...else
const score = 79
let grade;
if (score>90){
grade = 'A';
}
else if (score> 80){
grade = 'B';
}
else if (score>70){
grade = 'C';
}
else if (score>60){
grade = 'D';
}
else {
grade = 'F';
}
console.log(`Student's score is ${grade}`);
TERMINAL
→ ch4 git:(main) ✗ node conditionals.js
student's score is C
→ ch4 git:(main) ✗

Figure 4.24 – Conditional statements

Table 4.1 summarizes the most common conditional operators in JavaScript:

Operator	Description
==	Equal to
===	Equal value and equal type
!=	Not equal to
!===	Not equal value and not equal type
>	Greater than
<	Less than
>=	Greater than or equal to
<=	Less than or equal to

Table 4.1 – Conditional operators

In the next section, let us learn about JavaScript functions and how to use them to make the code more reusable.

Functions in JavaScript

Functions are used to package reusable code blocks to avoid redundancy. In JavaScript, the function is declared using the `function` keyword, followed by its name, parameters, and body. A function can accept zero or more parameters. Parameters must be separated by a comma if there is more than one. Functions can also return a value using the `return` statement followed by an expression or a value. Within the function, you can use the argument object to access the arguments as an array.

Figure 4.25 illustrates a function that computes the area based on the number of arguments sent:

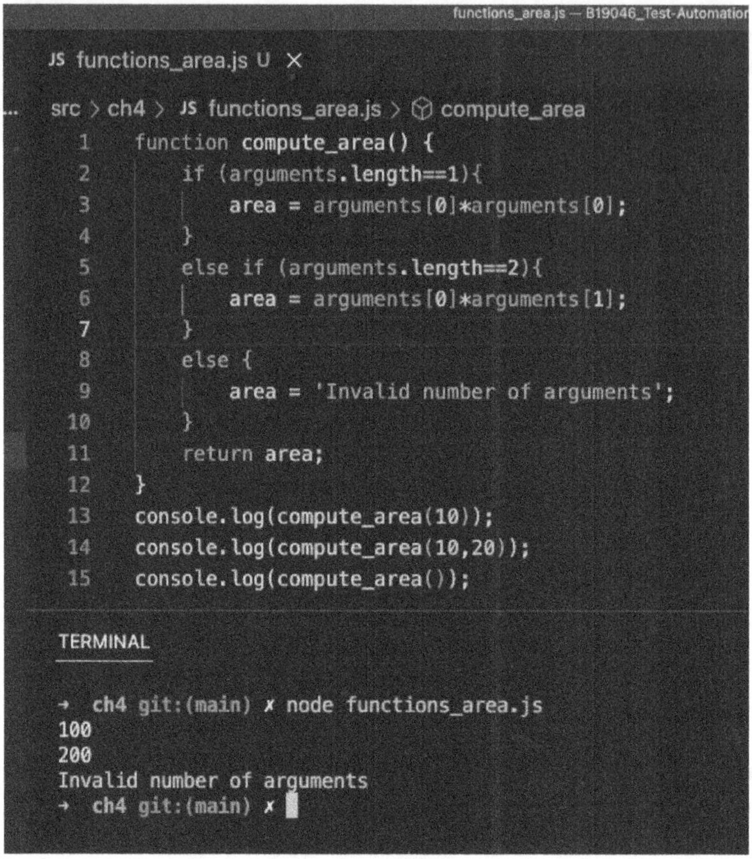

```
                                            functions_area.js — B19046_Test-Automation
  JS functions_area.js U ✕

... src > ch4 > JS functions_area.js > 🔶 compute_area
    1    function compute_area() {
    2        if (arguments.length==1){
    3            area = arguments[0]*arguments[0];
    4        }
    5        else if (arguments.length==2){
    6            area = arguments[0]*arguments[1];
    7        }
    8        else {
    9            area = 'Invalid number of arguments';
    10       }
    11       return area;
    12   }
    13   console.log(compute_area(10));
    14   console.log(compute_area(10,20));
    15   console.log(compute_area());

TERMINAL

→  ch4 git:(main) ✗ node functions_area.js
100
200
Invalid number of arguments
→  ch4 git:(main) ✗ ▌
```

Figure 4.25 – Functions

In the next section, let's look at how and where to continue our journey of learning JavaScript.

Exploring JavaScript further

The topics we have surveyed so far in JavaScript are a humble beginning and there are tons more to explore in this topic. ECMAScript 6 was a huge upgrade to JavaScript, and it has provided various efficient ways of performing routine programming tasks such as writing loops, defining functions, and more. We will be learning some of these more effective and newer ways in the subsequent chapters as needed. In the meantime, the MDN docs for JavaScript, located at https://developer. mozilla.org/en-US/docs/Web/JavaScript, can be used as a standard reference to get additional details on any concepts.

Summary

This chapter encompassed a considerable number of hands-on skills that a quality engineer uses daily. We started the chapter by looking into the various aspects of selecting an IDE. We moved on to downloading and setting up VS Code on macOS. We also learned how to search and use an extension from the VS Code Marketplace. We commenced our JavaScript learning in the subsequent section by understanding why we should learn it and laid a solid foundation with concepts such as variables and data types. We then learned about string operations before getting introduced to JavaScript objects. We dealt in detail with arrays and object literals.

We concluded our introduction to JavaScript by gaining knowledge about loops, conditionals, and functions. We grasped a little more detail on Git topics such as commits and merge conflicts. We closed out that section by learning some additional Git commands.

This sets us up excellently for the next chapter, *Test Automation for Web*. You will use your gained knowledge of JavaScript to automate a test using the Cypress tool. We will also look at the various considerations for web automation.

Questions

1. What are some factors to consider when choosing an IDE?

2. What is the difference between creating a variable using `let` and `const`?

3. How can we check whether an element is present in an array?

4. What are the advantages of object destructuring?

5. How are functions defined in JavaScript?

6. What does a `git bisect` command do?

7. What are some ways to avoid merge conflicts?

8. Why is a `git commit` message important?

5

Test Automation for Web

Test automation for modern web applications poses tough questions for quality engineers regularly, and there has been significant progress in terms of the tools and support in recent years. The test automation market is ripe with tools to aid engineers in setting up and executing efficient test scripts. Some notable tools include **Protractor**, **Cypress**, and **WebdriverIO**. The development of these tools with all their innovative capabilities is strong evidence of the evolution in the quality engineering space, primarily to address modern web automation demands. In this chapter, let us dive deep into automating a web application using Cypress and learn the nitty-gritty aspects of the tool.

Here are the main topics we will be looking at in this chapter:

- Why Cypress?
- Installing and setting up Cypress
- Creating your first test in Cypress
- Employing selectors and assertions
- Intercepting API calls
- Additional configurations
- Considerations for web automation

Technical requirements

In this chapter, we will continue using Node.js (version 16.14.2), which we installed in *Chapter 4, Getting Started with the Basics*. We will also be using **node package manager** (**npm**) to install Cypress version 11.2.0. All the code examples illustrated in this chapter can be found under the ch5 folder at `https://github.com/PacktPublishing/B19046_Test-Automation-Engineering-Handbook/tree/main/src/ch5`.

Why Cypress?

Cypress is a next-generation frontend test automation tool built for automating both legacy and modern web applications. It also comes loaded with additional power to perform in-depth component testing, mocking and stubbing API calls, and more. Our focus in this chapter will remain on exploring its **end-to-end** (**E2E**) testing abilities. Cypress has grown versatile in the space of web test automation as it comes packed with many handy tools that might need additional manual integration efforts when using some of the other tools in the market. Rather than limiting itself as a browser automation tool, Cypress comes as a comprehensive E2E testing framework to validate both backend and frontend logic. It has a similar configuration for every frontend framework, thereby making the setup easy and portable. It also has tremendous performance gains when compared with the test runtimes of other commonly used E2E test frameworks.

Several **continuous integration** (**CI**) services such as **CircleCI**, **GitHub Actions**, and more have specific modules that offer seamless integration with Cypress. Cypress also supports SDKs to scale the test execution across platforms on the cloud. There are also options to extend the behavior of Cypress using innumerable plugins, available at `https://docs.cypress.io/plugins/index`. Cypress thus brings enormous simplicity for its users in testing modern web applications, and in doing so, boosts their productivity.

To start learning Cypress, you should be aware of advanced software testing concepts. It is absolutely vital to possess reasonable experience working with JavaScript. It is recommended to go through the *Introduction to JavaScript* section from *Chapter 4* of this book to understand the basics. It would also help to have a good understanding of **object-oriented programming** (**OOP**) to assist with creating a robust test automation framework using Cypress. Knowing all that Cypress has to offer and more, it is a wonderful time to dive into the installation and setup of Cypress.

Installing and setting up Cypress

Let us now run through a detailed step-by-step installation and setup process for Cypress:

1. In *Chapter 4*, we installed Node.js, which is a runtime environment for JavaScript. Node.js installation comes with a default and extremely useful package manager called npm. *Figure 5.1* shows how to check the version of npm installed on your machine:

```
→  ch5 npm -v
8.15.0
→  ch5 
```

Figure 5.1 – Checking the installed npm version

2. Let us next create an empty project to install Cypress and further explore its features. Run npm init -y in an empty folder (preferably named app) in your local directory to create a package.json file. *Figure 5.2* shows the corresponding output with the contents of the file:

```
→  B19046_Test-Automation-Engineering-Handbook git:(main) cd src/ch5/app
→  app git:(main) npm init -y
Wrote to /Users/priya/Documents/workspace/B19046_Test-Automation-Engineering-Handbook/src/ch5/app/package.json:

{
  "name": "app",
  "version": "1.0.0",
  "description": "",
  "main": "index.js",
  "scripts": {
    "test": "echo \"Error: no test specified\" && exit 1"
  },
  "keywords": [],
  "author": "",
  "license": "ISC"
}


→  app git:(main) ✗ 
```

Figure 5.2 – npm init

> **Note**
>
> npm init <initializer> is used to set up new or existing packages. If <initializer> is omitted, it will create a package.json file with fields based on the existing dependencies in the project. The -y flag is used to skip the questionnaire.
>
> package.json is the primary configuration file for npm and can be found in the root directory of the project. It helps to run your application and handle all the dependencies.

3. Execute npm install cypress in the root of our src/ch5/app project. This creates a node_modules folder, which contains a chain of dependencies required by the package being installed (Cypress). *Figure 5.3* shows the output of this step, with package.json showing Cypress installed. It is generally considered good practice to save the testing libraries in the devDependencies section of the package.json file using the npm install cypress --save-dev command:

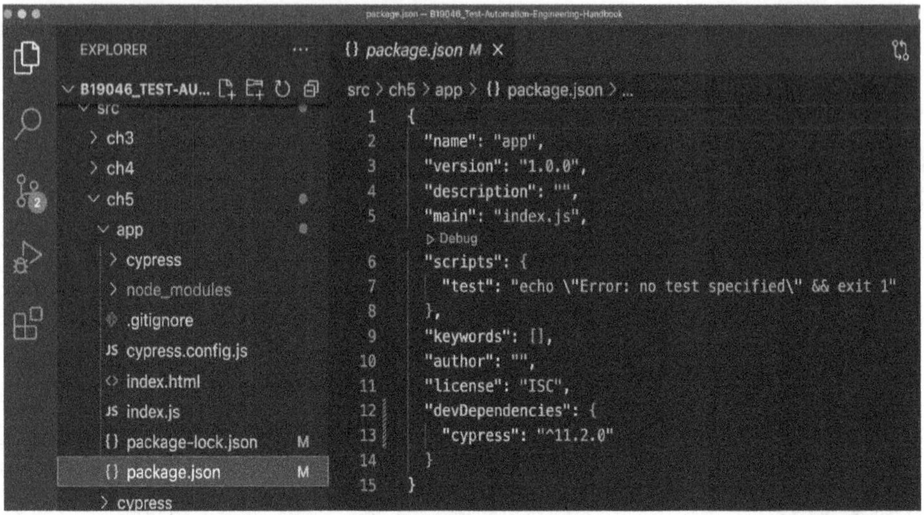

Figure 5.3 – npm install cypress

4.   Create an `index.html` file in the root, as shown in *Figure 5.4*, to serve as the primary loading page for our application. Also, create an empty `index.js` file:

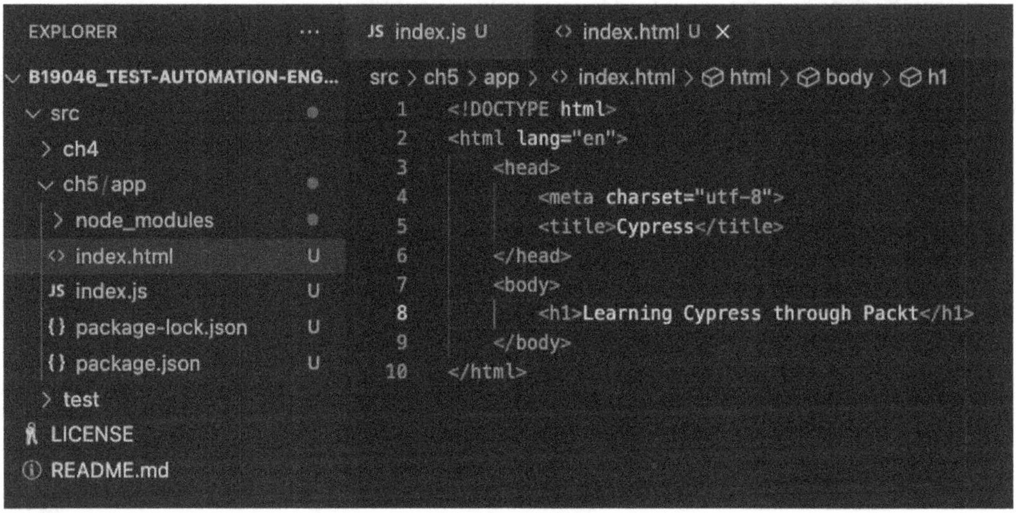

Figure 5.4 – Creating an index.html file

5. Execute npx cypress open to open Cypress. This command opens the executable file from the node_modules/bin directory. *Figure 5.5* illustrates the output where the in-built browser is opened:

Figure 5.5 – npx cypress open command

6. Now, click on the **E2E Testing** option, which adds some configuration files to the repository, and hit **Continue** in the next modal, as shown in *Figure 5.6*:

Configuration Files

We added the following files to your project:

cypress.config.js
The Cypress config file for E2E testing.

cypress/support/e2e.js
The support file that is bundled and loaded before each E2E spec.

cypress/support/commands.js
A support file that is useful for creating custom Cypress commands and overwriting existing ones.

cypress/fixtures/example.json
Added an example fixtures file/folder

Continue

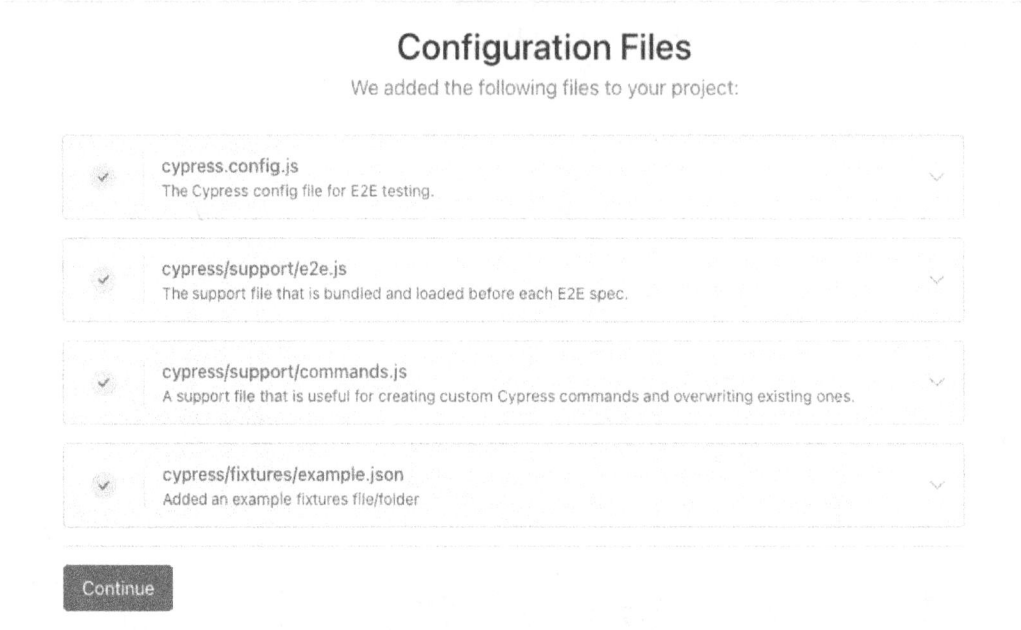

Figure 5.6 – Cypress config modal

7. In the next modal, select the preferred browser for E2E testing. I have selected **Chrome** in this
    case, as shown in *Figure 5.7*, and it opens the browser in a new window:

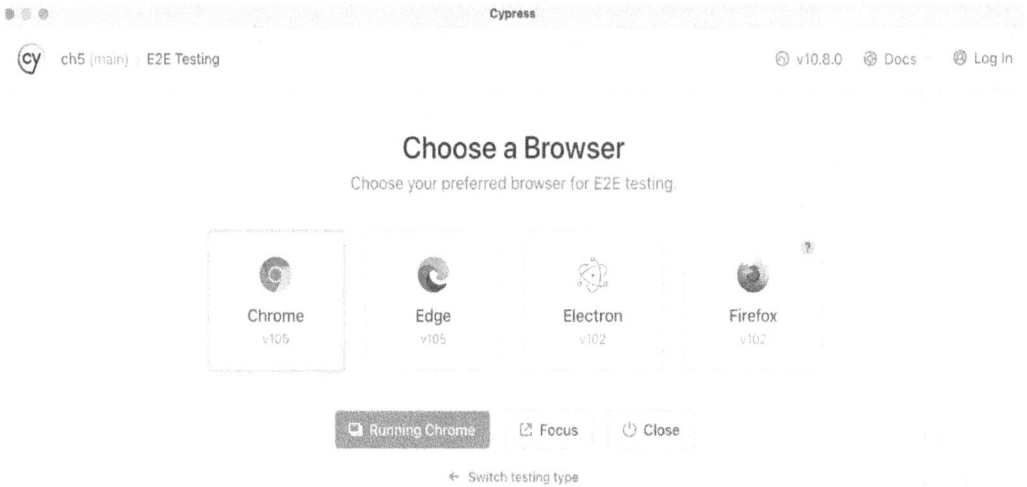

Cypress

(cy) ch5 (main) E2E Testing   v10.8.0   Docs   Log In

Choose a Browser

Choose your preferred browser for E2E testing.

Chrome   Edge   Electron   Firefox
v105   v105   v102   v102

Running Chrome   Focus   Close

← Switch testing type

Figure 5.7 – Choosing a preferred browser

This completes the installation of Cypress and gets it ready to a point where we can start writing our own tests. In the next section, let us start working on our first test and review some additional configurations.

Creating your first test in Cypress

A test in Cypress is commonly referred to as a **spec**, which stands for **specification**. We will be referring to them as specs for the remainder of this chapter. Let us begin by understanding how to write arrow functions and callback functions in JavaScript.

Creating arrow functions in JavaScript

Arrow functions are extremely handy, and they clean things up quite a bit. They were introduced in the **ECMAScript 6** (**ES6**) version. The code snippet in *Figure 5.8* shows a simple function to add two numbers. It takes two parameters and returns the sum. Let us turn this into an arrow function:

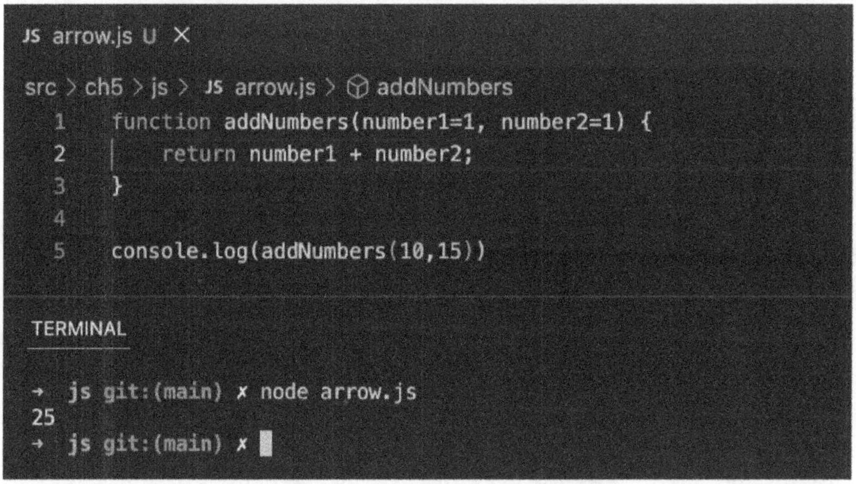

```
JS arrow.js U ×

src > ch5 > js > JS arrow.js > ⦾ addNumbers
  1    function addNumbers(number1=1, number2=1) {
  2        return number1 + number2;
  3    }
  4
  5    console.log(addNumbers(10,15))

TERMINAL

→  js git:(main) x node arrow.js
25
→  js git:(main) x ▮
```

Figure 5.8 – Function to add two numbers

Instead of using the `function` keyword, we name it like a variable and use an equals sign to assign it to the body of the function. After the parameters, we use a symbol called **fat arrow** (=>). In the case of one-liner functions, we can further simplify them by removing the curly braces surrounding the function body. We could also remove the `return` keyword, and it still returns the computed value. If we have only one parameter, we could lose the parentheses around the parameters as well. It would look like this: `const addNumbers = number1 => number1 + 5`. An example is shown in *Figure 5.9*. This works very neatly in the case of array iterations. Let's say we have an array of movies, and we would like to iterate over them and print the names of all the movies. This can be neatly done in a single line by using `movies.forEach(movie) => console.log(movie, name)` arrow functions:

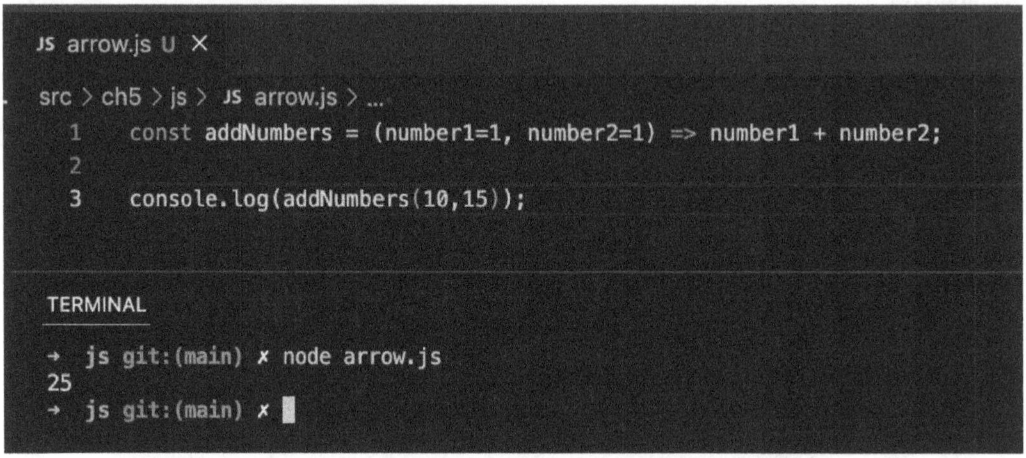

```
JS arrow.js U X
src > ch5 > js > JS arrow.js > ...
  1    const addNumbers = (number1=1, number2=1) => number1 + number2;
  2
  3    console.log(addNumbers(10,15));


TERMINAL

→  js git:(main) x node arrow.js
25
→  js git:(main) x █
```

Figure 5.9 – Arrow function with two parameters

Let us next learn about callback functions in JavaScript.

Creating callback functions in JavaScript

In JavaScript, since functions are set up as objects. we can pass and call other functions within a function. A function that is passed as a parameter to another function is called a `callback` function.

Let us use the `setTimeout()` function to understand callback functions. The `setTimeout()` function calls a method after a specified wait in milliseconds. For example, `setTimeout(() => console.log('hello!'), 5000)` would print the message after a wait of 5 seconds. Let us now create an arrow function to accept and print a message to the console, as shown in *Figure 5.10*. Let us call this function `printMessage()`, with a delay of 5 seconds by passing it as a parameter to the `setTimeout()` function, making it a callback function:

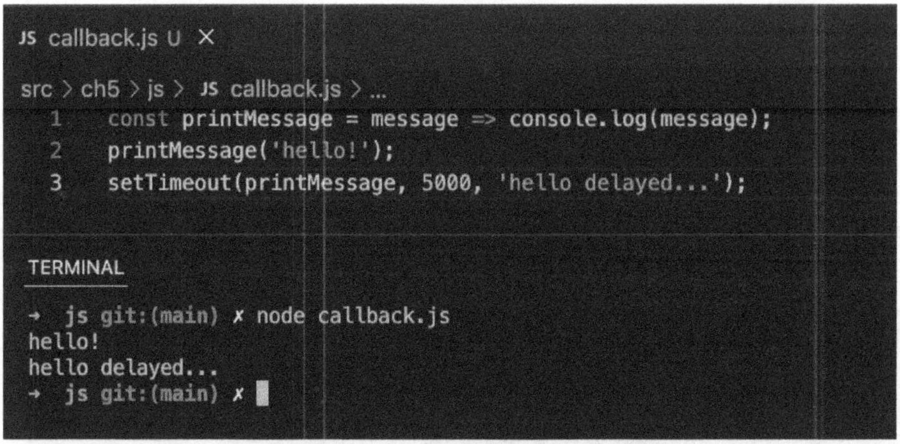

Figure 5.10 – Callback functions

We could also pass in the whole body of the arrow function instead of the name, as shown in *Figure 5.11*. These are called anonymous functions since they do not have a name and are declared at runtime:

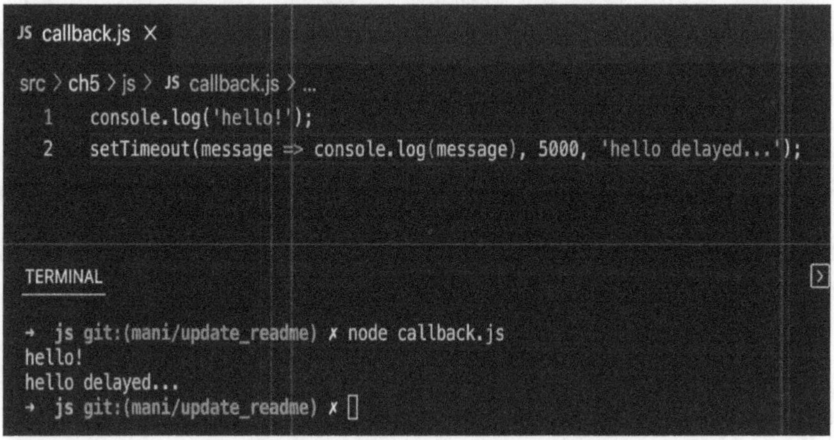

Figure 5.11 – Anonymous callback functions

A key advantage of using callback functions is that it enables the timing of function calls and assists in writing and testing asynchronous JavaScript code. There are many instances in the modern web application where there is a need to make external API calls and resume the current task rather than wait for the external call to complete. Callback functions come in handy here to unblock the execution of the main block of code. It is important to use callbacks only when there is a need to bind your function call to something potentially blocking, to facilitate asynchronous code execution.

With this additional knowledge about functions in JavaScript, let us now commence writing our first spec.

Writing our first spec

It is a good practice to organize all tests under a single folder in your repository. If there are more tests, then they can be categorized under a parent test folder. Create a folder named e2e under `src/ch5/app/cypress`. Now, create a test file, as shown in *Figure 5.12*:

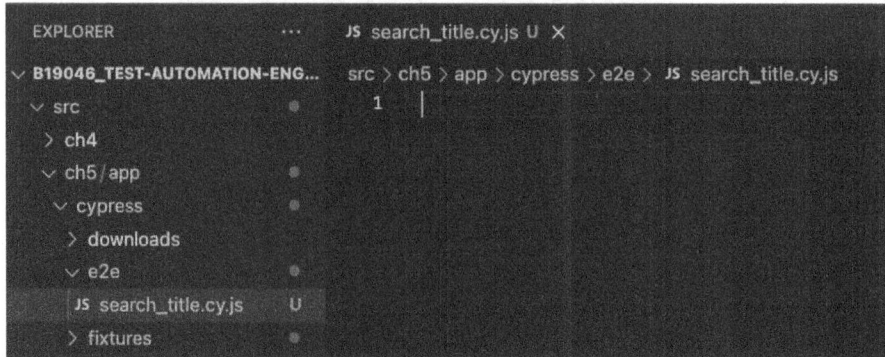

Figure 5.12 – Creating a test file

Our first test searches for the string `quality` in the search box on the home page of `https://www.packtpub.com/`. Then, it verifies the search result page by looking for the `Filter Results` string. Copy and paste the code from the `https://github.com/PacktPublishing/B19046_Test-Automation-Engineering-Handbook/blob/main/src/ch5/app/cypress/e2e/search_title.cy.js` GitHub link into the test file.

Let us now examine the structure of a Cypress spec.

Becoming familiar with the spec structure

Every test framework requires its tests to be written in a specific language and format. Cypress is no different, and as we already know, it uses JavaScript. Cypress comes packed with its own set of functions under the global cy object. It also utilizes the *describe-it-expect* format using bundled libraries from **Mocha** and **Chai** frameworks. Additionally, an assertions framework using `expect` with command chaining is also supported to complete granular validations. The `describe` block captures the high-level purpose of the spec, and the `it` block adds specific implementation details of the test. Note that both the `describe` and `it` blocks accept callback functions as their second parameter, and they are defined as arrow functions. This is a common syntax, and you will see this more often in modern JavaScript code. Please be wary of braces, semicolons, and parentheses. It is recommended to use an extension such as **Prettier** to assist with the formatting as it could get messy pretty quickly.

We have started with a comment that describes what is being achieved in this spec. Cypress internally uses the **TypeScript** compiler, and the reference tag is used to equip autocompletion with only Cypress definitions. The `beforeEach` block, as the name suggests, runs before every `it` block. It usually contains the prerequisite steps to execute the individual `it` blocks. Here, we use the `visit` command to access the *Packt Publishing* website within the `beforeEach` block. Then, the `it` block drills down to which action is performed in the spec. If we end up adding more `it` blocks to this spec, the `visit` command would be executed before the beginning of each `it` block. This is a simple spec but it captures the necessary structure of a spec written in Cypress.

Next, let us examine how to execute our first spec.

Executing our first spec

Cypress comes packed with a powerful visual runner tool to assist in test execution. This can be used when users have a need to inspect tests visually during runtime. Another option is to execute tests via the CLI for quicker results and minimal test execution logs. In this section, we will survey both ways to execute tests in Cypress.

Using the command line

Using the command line to execute tests is always a quick and easy option. It usually helps when you are not interested in looking at the frontend aspects of the test execution. The `npx cypress run -s cypress/e2e/search_title.cy.js` command can be used to execute an individual spec in Cypress. The `-s` flag stands for spec, followed by the name of the file. Without the `-s` flag, the `npx cypress run` command would execute all the specs found in the current project. *Figures 5.13* and *5.14* illustrate the output of the command-line execution of our first spec. *Figure 5.13* shows the output of the CLI, with a listing of actions performed on the UI:

```
→  app git:(main) ✗ npx cypress run -s cypress/e2e/search_title.cy.js


(Run Starting)

 ┌─────────────────────────────────────────────────────────────────────
 │ Cypress:        11.2.0
 │ Browser:        Electron 106 (headless)
 │ Node Version:   v18.12.1 (/usr/local/bin/node)
 │ Specs:          1 found (search_title.cy.js)
 │ Searched:       cypress/e2e/search_title.cy.js
 └─────────────────────────────────────────────────────────────────────


──────────────────────────────────────────────────────────────────────

 Running:  search_title.cy.js                                        (1
of 1)


 Vist packt home page,
   ✓ search for a title and and click submit (11003ms)
   ✓ by intercepting the recent items GET call (1802ms)


 2 passing (14s)
```

Figure 5.13 – CLI test execution

Figure 5.14 shows a summary of the tests executed, with a breakdown of the results:

```
═══════════════════════════════════════════════════════════════════════

 (Run Finished)


       Spec                                Tests  Passing  Failing  Pending  Skip
ped
 ┌─────────────────────────────────────────────────────────────────────
 │ ✓  search_title.cy.js          00:13     2       2        -        -        -
 └─────────────────────────────────────────────────────────────────────
   ✓  All specs passed!           00:13     2       2        -        -        -

→  app git:(main) ✗ █
```

Figure 5.14 – CLI test execution (continued)

Next, let us next explore the visual test runner for executing our spec.

Using the visual test runner

Cypress comes with an extremely insightful and detailed test runner and provides quite a bit of debugging data for tests being executed. To utilize this mode, we can start with the `npx cypress open` command, which opens up a series of modals. The first modal requires the selection of the type of test, as previously shown in *Figure 5.5*. The second modal, as seen earlier in *Figure 5.7*, provides an option to select a browser against which the test can be run. The third modal lists all specs in the project and shows some additional metadata about the specs and their recent runs:

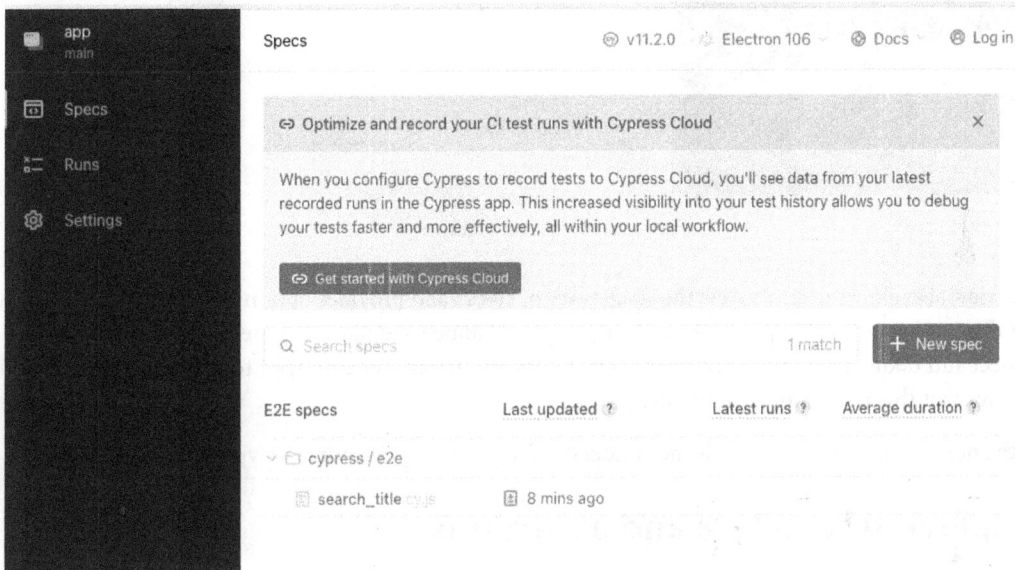

Figure 5.15 – Test selection modal

Test execution begins when the user clicks on the test, as shown in *Figure 5.15*. This opens a new browser that shows the actual steps being executed. The left pane shows the various frontend and backend calls being made while executing this test. *Figure 5.16* shows a view of the test execution. Cypress offers a live-reloading feature out of the box using the `cypress-watch-and-reload` package. Whenever there is a change in the code, the test is rerun automatically and the view, as shown in *Figure 5.16*, reloads live:

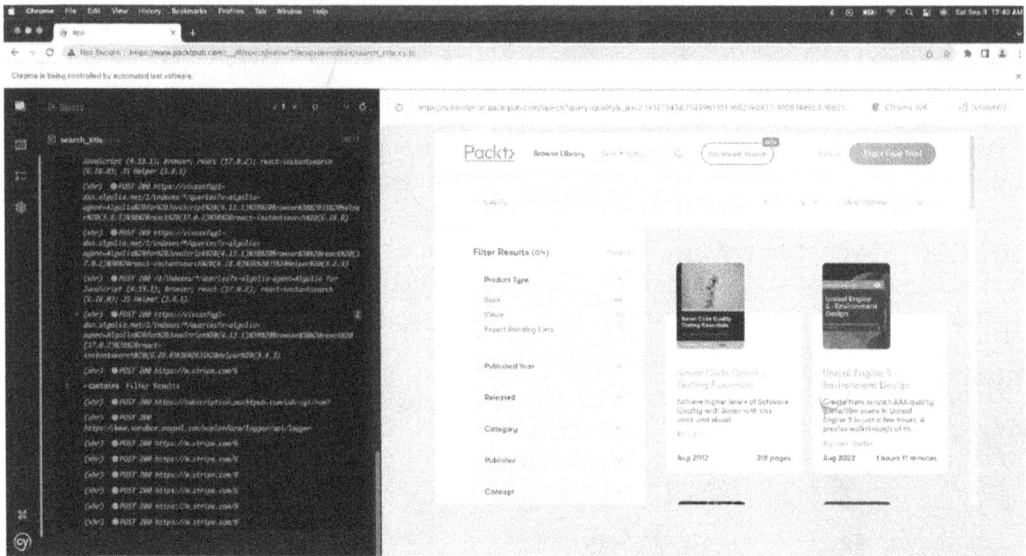

Figure 5.16 – Visual test runner

This view also allows users to view the stack trace of errors and provides options to navigate between test runs and settings. The browser on the right pane can be used like any other browser window to inspect and debug using the developer tools. Users are strongly encouraged to further explore the features that this test runner has to offer.

In the next section, let us gain a deeper understanding of using selectors in Cypress.

Employing selectors and assertions

Selectors are identifiers for elements in the **Document Object Model** (**DOM**). We have various ways to identify elements, such as using their class, name, type, and so on. Every test framework has its own custom commands to make the code clear and concise. Cypress provides users with an efficient interface to look for selectors and comes with standard support for all selectors. Let us continue using our first spec to dig deeper into utilizing selectors.

Working with selectors

`cy.get` is the primary function to search for elements in the DOM. In our `search_title.cy.js` test file, we have used `.input-text`, which identifies the element with the `input-text` class name and sets a value in it. We have also used `[aria-label="Search"]` to look for the **Search** button. This is an example of an attribute search. We are essentially finding an element with the value of the `aria-label Search` attribute and clicking on it. `id` and `data` are other reliable attributes for identifying elements in the DOM. It is important to remember to use square brackets when employing attributes in selectors. This raises the question of what kind of selector to use in each case. The answer would be to employ the simplest one that uniquely identifies the required element on the DOM.

Cypress assists users here by providing a selector playground feature that automatically populates the selector. Let us rerun our first spec using a visual test runner and reach the execution page, as shown in *Figure 5.16*. Now, refer to *Figure 5.17* and click the circular toggle icon This opens the selector playground where the user can type the selector or use the arrow icon for Cypress to automatically populate it. Now, the user can use the browser to click on the required UI element and get the unique selector right away. The user can also play around with other options and validate their correctness by plugging them into the textbox:

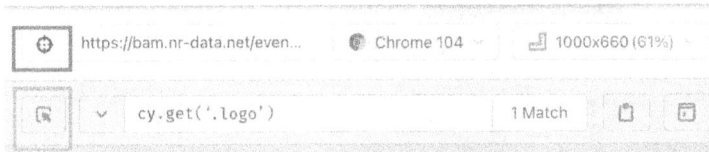

Figure 5.17 – Selecting a playground

To write efficient automation scripts, it is vital to know which selectors are reliable and perform better in a given situation. Imagine a test automation project with 5,000 test cases and all of them find a link using the worst-performing selector, which has a lag of 50 milliseconds relative to the best-performing selector. That would make the test suite slower by 250,000 milliseconds for every run. This would impact feedback times immensely when considering hundreds of CI pipeline runs over a few days.

XPath selectors identify the target element by navigating through the structure of the HTML document, while CSS selectors use a string to identify them. Direct CSS selectors using an element's ID or class usually perform better than XPath selectors. Using an ID selector is often the most reliable way of selecting an HTML element. It helps to analyze the elements to understand whether they are dynamic and which selectors would be supported across different browsers, and based on that, decide on a selector strategy. It usually takes a bit of troubleshooting to arrive at an efficient pattern of selectors working for a specific application and a team.

Let us now learn about the available assertion options.

Asserting on selectors

Assertions give us a way to perform validations on various UI elements. They are usually chained to the command with selectors and work together to identify an element and verify it. `should` is the primary function utilized on assertions, and it works with a myriad of arguments.

Let us update our first spec to add some assertions. We have earlier used the `contains` function in our spec to assert a partial string in the search results page. *Figure 5.18* shows the assertions in action. Next, we add an assertion on the **Reset** button to validate that it is disabled. In the following line, we get the navigation bar element by the `id` attribute and chain it with an assertion that validates the class name:

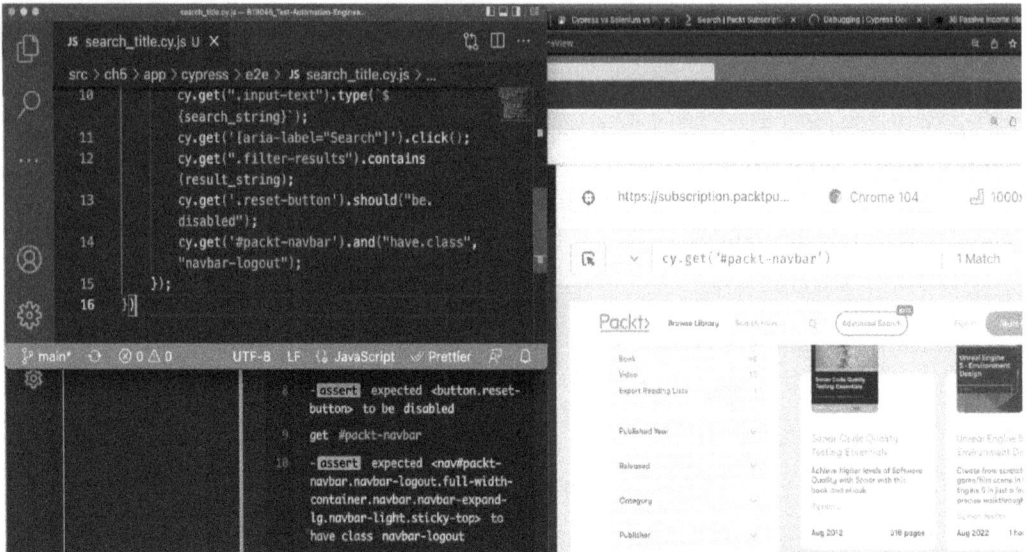

Figure 5.18 – Assertions for the navbar and Reset button

Let us add another assertion before entering the search string to validate it is empty using the `have.value` parameter. *Figure 5.19* demonstrates this assertion:

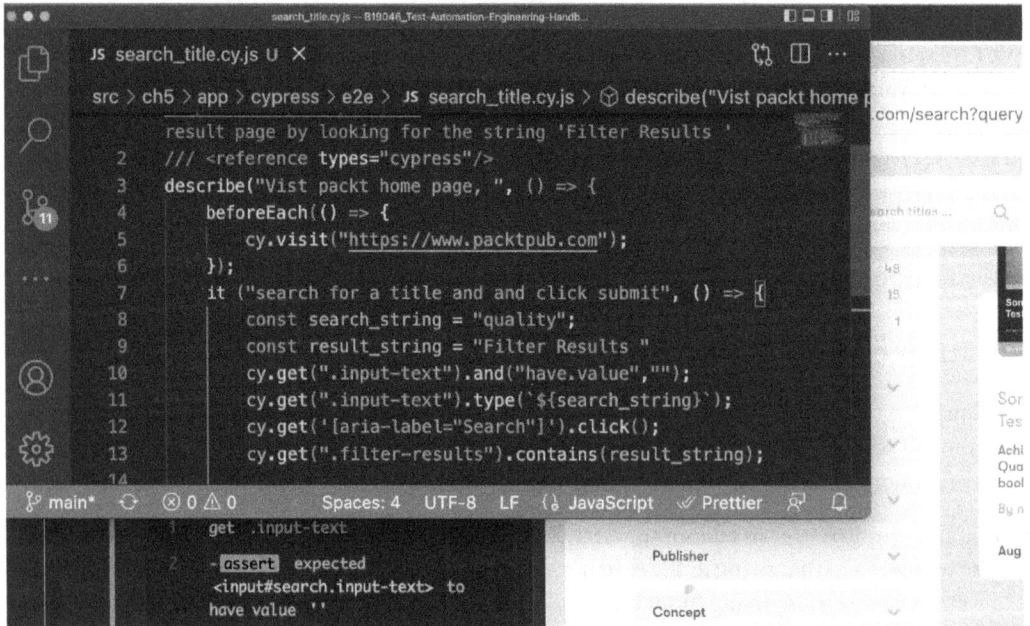

Figure 5.19 – Asserting empty value in a textbox

Cypress comes with very good documentation (`https://docs.cypress.io/api/table-of-contents`) and users are encouraged to use it as a reference to be aware of the various available options. So far, we've worked on identifying and asserting UI elements. In the next section, let us work with API calls in Cypress.

Intercepting API calls

Cypress lets users work with underlying API requests and stub responses where necessary. Let us analyze the API calls when loading the *Packt Publishing* home page and try to stub one of the responses. `cy.intercept()` is the command used to work with API calls, and it offers a wide variety of parameters. For this example, we will be using the `routeMatcher` and `staticResponse` arguments. We add a second `it` block to intercept the underlying API call and specify the type of HTTP call, URL, and a predefined response as parameters, as shown in *Figure 5.20*:

```
40        });
41        cy.intercept(
42          "GET",
43          "https://subscription.packtpub.com/api/subscription/getrecentitems?offset=0&limit=4",
44          staticResponse
45        );
```

Figure 5.20 – cy.intercept call

The value of the static response parameter can be obtained using the **Network** tab of the developer tools to get the actual response for the API call. This is illustrated in *Figure 5.21*. By passing this in as the `staticResponse` parameter, the `GET` call on this URL will always return the stubbed response instead of the original:

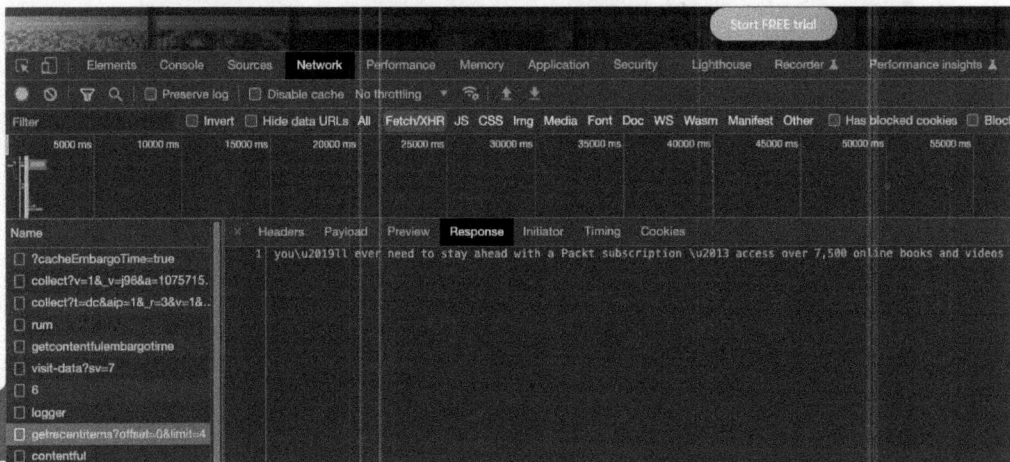

Figure 5.21 – API payload for stubbing

Figure 5.22 demonstrates the result of the intercept command in action:

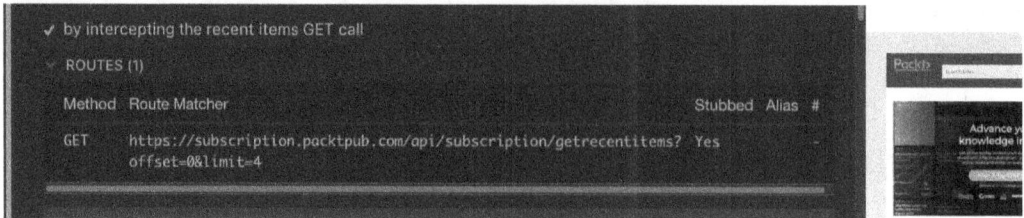

Figure 5.22 – Intercept results

This empowers the user to test the underlying API calls for different payloads and validate the application behavior in each case. This also saves resources in cases where some of these are expensive API calls. This is just one way to handle API calls with Cypress, and there are a variety of options available to explore. In the next section, let us quickly review some additional configurations that might be helpful with setup and validation.

Additional configurations

Let us review a few additional configurations in this section to build stable and efficient specs:

- The first configuration is a Git feature and not specific to Cypress. Adding a .gitignore file is a general necessity for all projects. Add it to the src/ch5/app root folder to ignore files we don't want Git to track on our local directory. As shown in *Figure 5.23*, let's add a node_modules folder so that we don't have to check in and keep track of all dependencies:

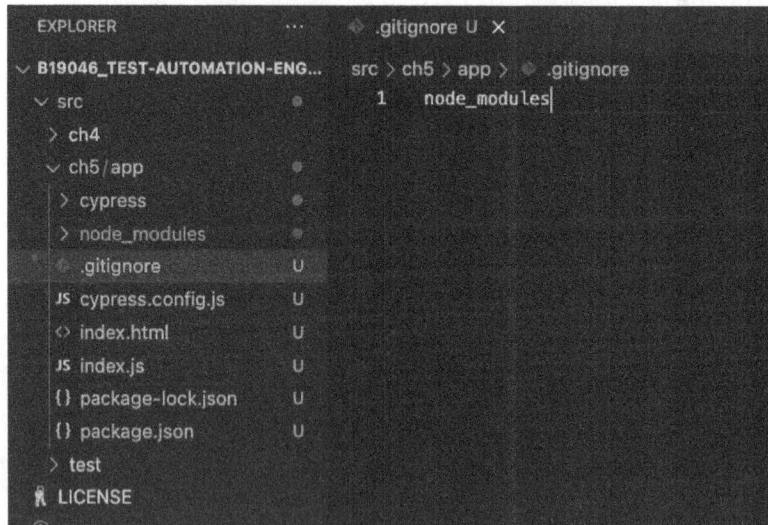

Figure 5.23 – .gitignore file

- Cypress comes with a default timeout of 4 seconds (4,000 milliseconds), and the `get` method accepts a second parameter to set a custom timeout. For example, in our spec, we can add extra wait after searching for the string and waiting for the **Reset** button to appear with `cy.get('.reset-button', {timeout:10000}).should("be.disabled")`. This line waits for 10 seconds for the **Reset** button to appear and then runs the assertion.

- Cypress provides a delay feature when performing actions on the DOM. For example, when typing an entry in the textbox, it has a default delay of 10 milliseconds for each key press. In our spec, this delay can be customized as `cy.get(".input-text").type(`${search_string}`, { delay:500 })` to fit the need of the application. In this case, there will be a half-second delay before typing the next character in the textbox.

We have secured a solid understanding of the major features of Cypress and are well set to explore its capabilities further. Before we close this chapter, let us review some valuable considerations for web automation.

Considerations for web automation

Modern web application testing and automation come with their own set of challenges in terms of complexity. With all the great features packed within Cypress, it has its own limitations as well.

Limitations of Cypress

So far, we have reviewed in detail what Cypress can do, but it would be wise to also call out the things Cypress cannot do as of today. When considering test automation at an organizational scale, it is critical to gather clarity on how the tool fits in. Let us now list a few items that Cypress does not support as of today:

- It currently cannot drive two browsers in concurrency.
- It doesn't provide support for multi-tabs in the browser.
- It only supports JavaScript for test cases.
- It currently does not support Safari or IE.
- It has limited support for iFrames (an element that loads another HTML element within a web page).
- The larger the test suite, the slower its execution in Cypress due to its underlying protocol, which enables a debugger in Chromium-based browsers. It is, however, ideal for small-to-medium-sized test suites.
- Cypress contains a lot of syntactic ambiguity that can make it difficult to scale it across an organization.

Let us now take up some considerations for web automation in general.

Web automation considerations

So, it is imperative to thoroughly review your team's test automation requirements and take into consideration every detail. Let us look at some chief items to be mindful of:

- Always focus on validating the core business logic and the scenarios that surround it. Moving past the basic functionality and building a stable automated regression suite for your web application gives tremendous ROI in the long term.

- In the era of digitization, the usability of the web application takes a focal role, and it is compulsory to validate the look and feel of the web application across multiple browsers on multiple platforms.

- It is essential to perform due diligence regarding continued support from the community or the vendors before deciding on a tool or strategy for web application automation.

- We can't stress enough the critical role Agile methodologies play in modern software projects. So, it is vital for the web application tool to integrate seamlessly with CI services and perform efficiently in union.

- The data and privacy aspects of test data should be taken into account since modern web applications tend to tap the power of cloud services continually. Further attention has to be paid when handling **personally identifiable information (PII)**.

- To gain additional stability with test scripts, avoid cluttering them with brittle selectors. Exercise caution when choosing a selector and have a long-term view of how the screen might change.

- Distinguish between synchronous and asynchronous calls within your web application and employ test methods appropriately to suit the purpose.

- Make efforts to organize tests consistently for readability and performance reasons. Explore the tool's capabilities or additional plugins to standardize test scripts.

This brings us to the end of a detailed introductory walk-through of Cypress and its features. There is more to explore and learn, as with any other tool. Cypress offers great promise for web application automation and provides innovative solutions to some of the problems that have haunted the test engineering community in modern web application testing. The Cypress team actively releases a lot of new features on a regular basis, and it is the right time to further explore its advanced abilities. Let us summarize what we have learned in this chapter in the next section.

Summary

Let us quickly recap what we have learned in this chapter. We went through a step-by-step installation and setup process for Cypress. We commenced the next section by understanding arrow and callback functions in JavaScript. Then, we continued on to write our first spec and ventured out to comprehend its structure and execution capabilities. We worked on using selectors and assertions within our spec to identify and validate DOM elements. Then, we learned how to intercept API calls in Cypress. We familiarized ourselves with some additional Cypress configurations before taking on web automation considerations.

In the next chapter, we will confront another crucial quality engineering topic: mobile test automation. We will be using Appium 2.0 to formulate test scripts for Android and iOS platforms.

Questions

1. What are some advantages of using Cypress?

2. What is the purpose of the `package.json` file in the code repository?

3. Why are callback functions used in JavaScript?

4. What is the purpose of a `beforeEach` block in a Cypress spec?

5. How do you select a DOM element using the `id` attribute?

6. What is the Cypress command used to intercept API calls?

6

Test Automation for Mobile

The demand for releasing quality mobile applications across multiple platforms warrants a solid test automation framework to thoroughly inspect and catch any regression issues. Mobile application developers have the necessity to rapidly release bug fixes and new features to keep up with the competition. This makes mobile test automation indispensable in the development life cycle. In this chapter, let us take a hands-on approach to thoroughly review mobile test automation and its setup.

Here are the main topics we will be looking at in this chapter:

- Getting to know Appium

- Knowing WebdriverIO and its advantages

- Setting up Appium and WebdriverIO

- Writing our first mobile test

- Key considerations for mobile automation

- Optimizing our mobile automation framework

Technical requirements

We will continue using Node.js, which we installed in *Chapter 4*. We need the **Java Development Kit (JDK)** on our machines (which can be found at this link: `https://adoptium.net/temurin/releases`) to download and install the compatible version with the operating system. Please remember to set up the JAVA_HOME path in your `.zshrc` or `.bashrc` files. JAVA_HOME should point to the directory where the JDK was installed.

Next, we will be needing Android Studio, which can be installed using the following link: `https://developer.android.com/studio`. Similar to the JAVA_HOME path, the ANDROID_HOME variable should be set up in your `.zshrc` or `.bashrc` files and point to the directory where the Android SDK is installed. Also, make sure to append the PATH variable to include the `platform-tools` and `tools` folders within the SDK. We will be using a demo Android application from `github.com/appium/android-apidemos` for our automated testing. This is Google's demo application used for testing Appium. All the code and setup used in this chapter can be found at `https://github.com/PacktPublishing/B19046_Test-Automation-Engineering-Handbook/tree/main/src/ch6`.

Getting to know Appium

Appium is one of the most popular test automation frameworks on the market. It has several salient features that help automate iOS and Android applications. Let us start with knowing its fundamentals.

What is Appium?

Appium is an open source tool for automating native, hybrid, and mobile web applications. Appium primarily operates on iOS/Android mobile applications and can support Windows desktop applications as well. Appium uses vendor-provided frameworks under the hood—namely, **XCUITest** and **UIAutomator** for iOS and Android platforms, respectively.

Appium wraps these vendor-provided frameworks into the WebDriver API. As we have seen in *Chapter 2*, Appium uses a client-server architecture and extends the existing WebDriver protocol by adding additional API methods to support mobile automation. For engineers who have previously worked with WebDriver for web automation, this should be quite familiar and should not be brand new. It supports multiple programming languages such as Java, JavaScript, Python, Ruby, and so on.

Now, let us review Appium's advantages in the next section.

Advantages of using Appium

One of the chief advantages of Appium is its *cross-platform compatibility*. It allows users to write tests against multiple platforms using a single API. Configurations for both Android and iOS can be done within the same project. This enables the reuse of the majority of the code between iOS and Android platforms with minimal configuration changes. Since it uses vendor-provided frameworks, we do not have to compile Appium-specific code or third-party integrations as part of the application being tested. It can also work with major test runners and frameworks out there due to its compatibility with a wide variety of programming languages. Another primary feature of Appium is that it supports writing tests on both emulators and physical devices. Being an open source tool, it has tremendous community support, and anyone can extend its attributes.

Let us next look into what WebdriverIO is and review its advantages.

Knowing WebdriverIO and its advantages

WebdriverIO is an **end-to-end** (**E2E**) test automation framework written in JavaScript that lets you automate modern web applications in different browsers and operating systems. With the help of Appium, it also supports automating mobile applications in iOS and Android. WebdriverIO is simple and quick to get started with very few lines of code.

It provides simple wrapper methods to interact with web elements that are easy to pick up for beginners and, more importantly, helps write clean code. Since frontend development happens more commonly in JavaScript, it is possible for all engineers to easily contribute to tests. It is also a strong open source tool with huge community support.

Having looked at what Appium and WebdriverIO are, let us now go over the steps to install and set them up.

Setting up Appium and WebdriverIO

Combining Appium with WebdriverIO helps build an extremely scalable and customizable mobile automation framework. Let us start with Appium installation on macOS.

Appium installation

Next, let us go ahead and install Appium. This can be done using the `npm install -g appium@next` command to install a version above 2.0 globally. The use of `@next` here is to force an installation of a beta version (2.0.0). If by the time you install this, 2.0 is available as the latest stable version, you do not have to use `@next`. With the `appium -v` command, double-check the version after installation.

Appium involves multiple installations working in tandem, and it would be convenient to have a tool that could provide us with constant feedback on the health of our setup. `appium-doctor` does just that. Let us install it using the `npm install -g appium-doctor` command, and the installation can be checked using `appium -version`.

`appium-doctor` can be executed for the Android platform through the `appium-doctor --android` command. It primarily checks for the presence of all dependencies and their path. Users should be on the lookout for any errors in the output from executing `appium-doctor` and fix them before proceeding further.

There is one more installation before we can wrap up this section. Let us now install the necessary Appium drivers for iOS and Android using the following commands:

```
appium driver install xcuitest
appium driver install uiautomator2
```

The installations can be verified using `appium driver list`, as shown in *Figure 6.1*:

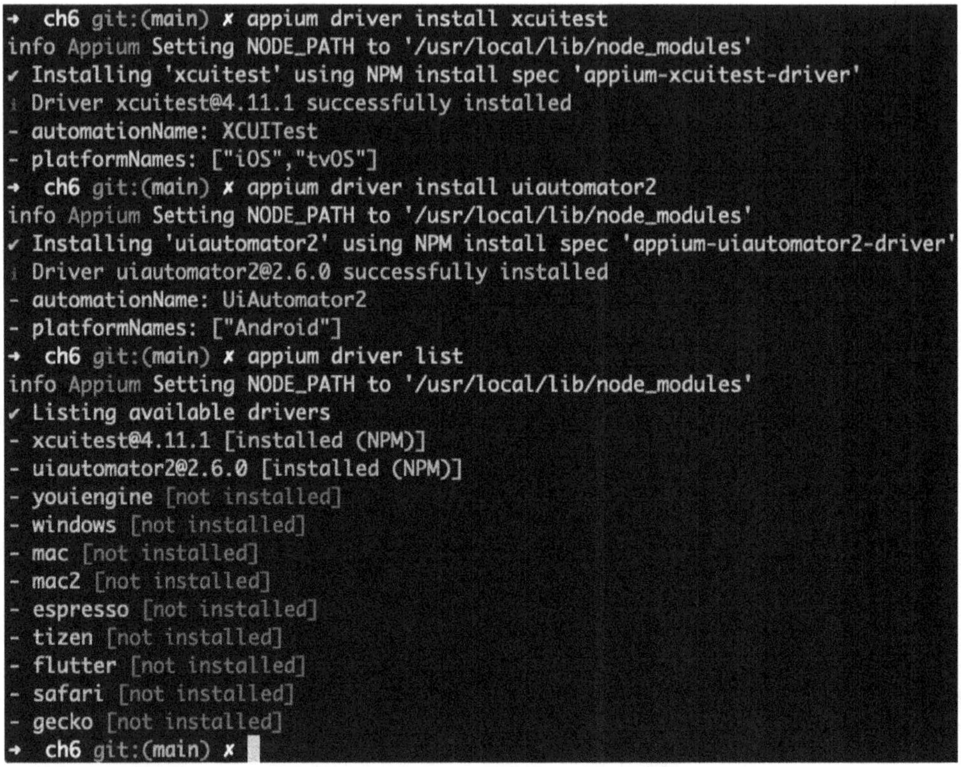

Figure 6.1 – Appium driver installation

This completes the Appium setup for macOS, and we have a couple more steps before we can start writing our first tests. Before diving into the WebdriverIO setup, let us look at how to configure an Android emulator.

Configuring an Android emulator

Emulators are virtual devices that let engineers set up and test against a variety of devices on the computer. Let us go over how to set up an Android emulator in this section:

1. In the home dialog of Android Studio, click on the **Virtual Device Manager** option, as shown in *Figure 6.2*:

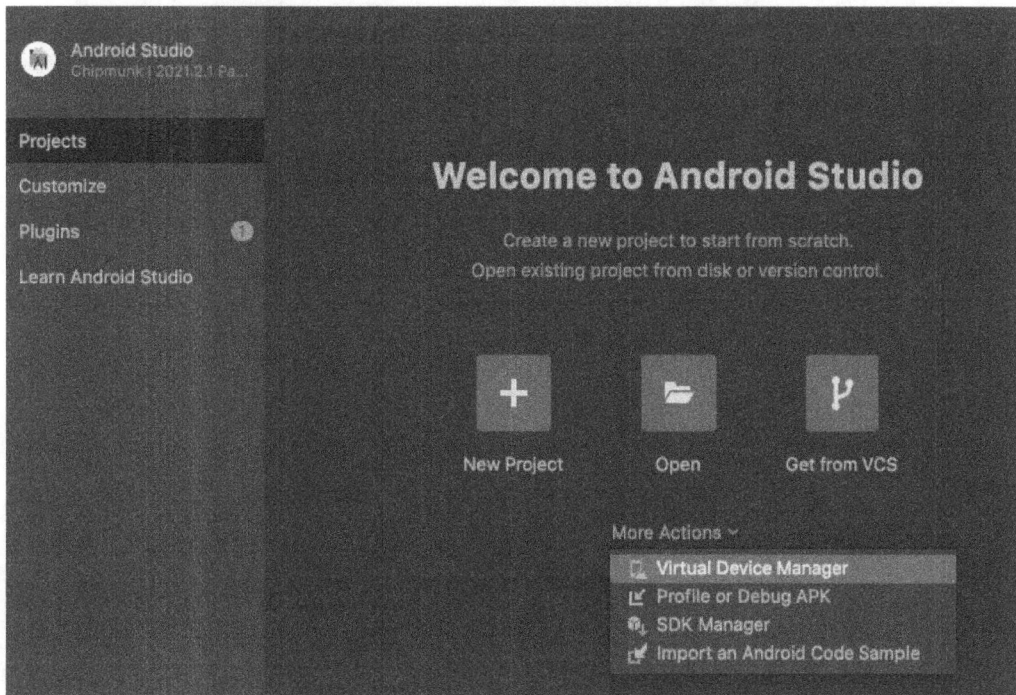

Figure 6.2 – Android Studio Virtual Device Manager

2. The Android Studio installation usually comes with an emulator, as shown in *Figure 6.3*. You can add another device by clicking on the **Create device** button at the top left of the dialog box. This opens an additional dialog to set up the new hardware:

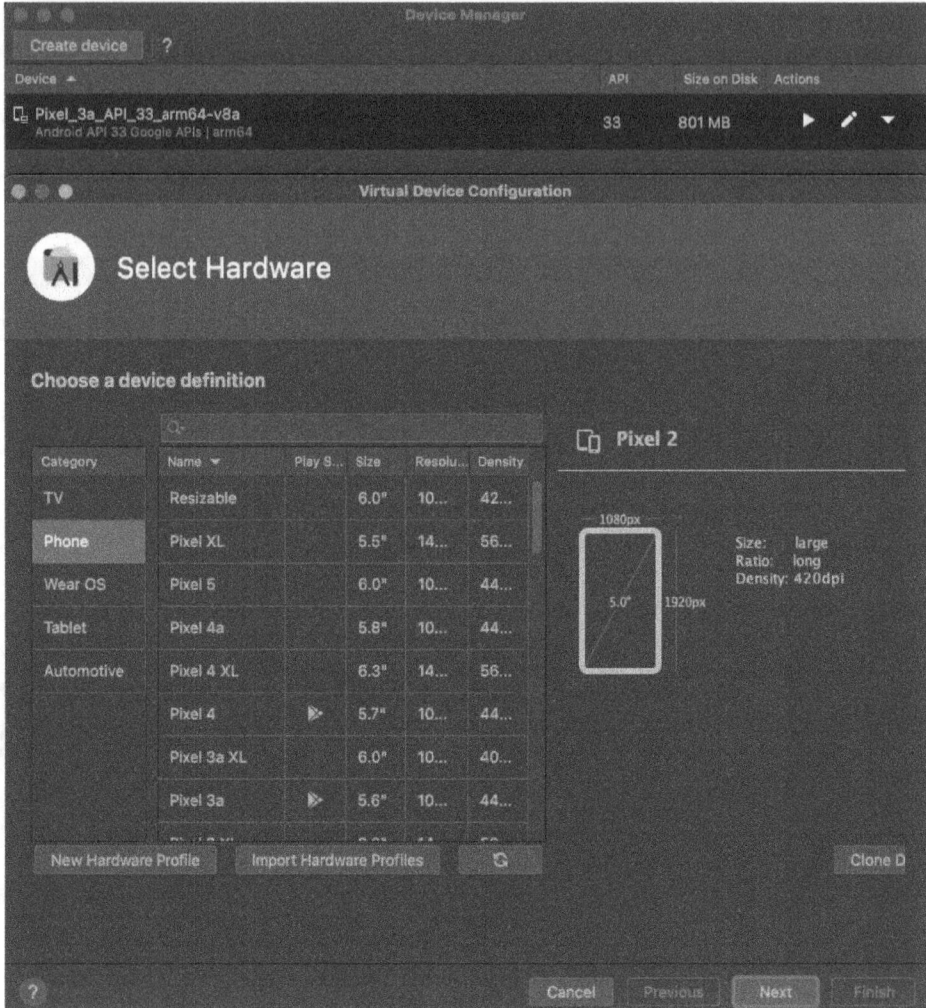

Figure 6.3 – Android Studio device selection

3.   For our example, let us select **Pixel 4a** and hit **Next**, which will take us to the **System Image** selection dialog.

4.   Please make sure to download two different versions here. We will be running the Appium tests on one version and connecting the Appium Inspector on the other. In our case, we will be downloading the Android API versions 32 and 33, as shown in *Figure 6.4*:

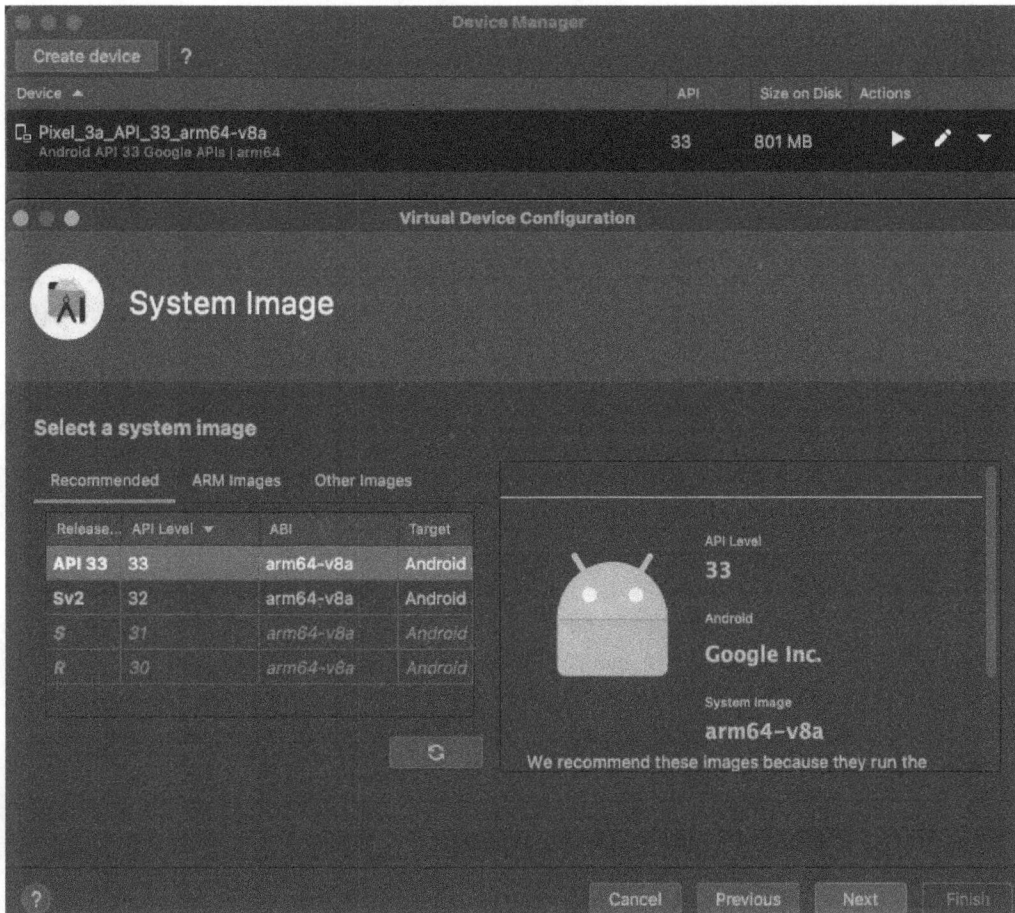

Figure 6.4 – Android Studio system images

5. Hit the **Next** button and complete the emulator setup with the **Finish** button.

6. On the **Device Manager** dialog, hit the play button against one of the devices, as highlighted in green in *Figure 6.5*. This will open the device emulator:

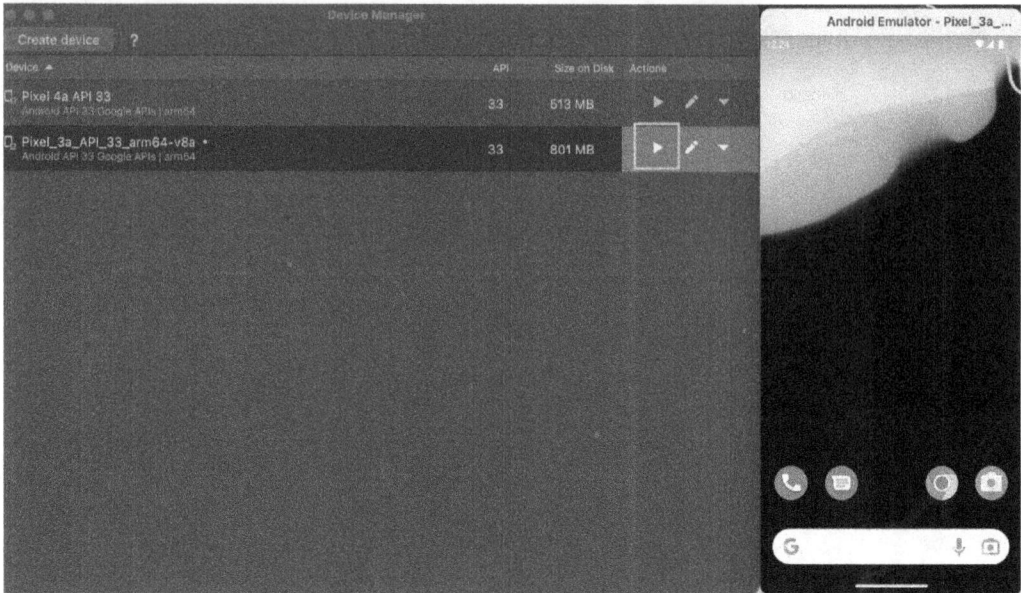

Figure 6.5 – Opening the Android device emulator

We are now set up with Appium and an emulated device to test on. Let us now dive into WebdriverIO installation and setup.

Configuring WebdriverIO with Appium

Let us begin with the WebdriverIO setup by creating and initializing a new directory named `webdriverio-appium`. You could also name it based on whichever project name suits your needs. Once we are inside the directory, let us run the `npm init -y` command to set up our project. This should create a `package.json` file within the directory.

Next, let us install the CLI for WebdriverIO using the `npm install @wdio/cli` command. This is used for setting up the configuration for WebdriverIO using the CLI. Let us now run `npx wdio config`. This `npx` command goes inside the `node_modules` folder to find WebdriverIO, which we just installed, and runs the `config` command using the CLI.

On running this command, we go through a series of steps, with each of them requiring a selection. Let us look at them one by one. Refer to *Figure 6.6* for a summary of the options selected for configuring WebdriverIO through the CLI:

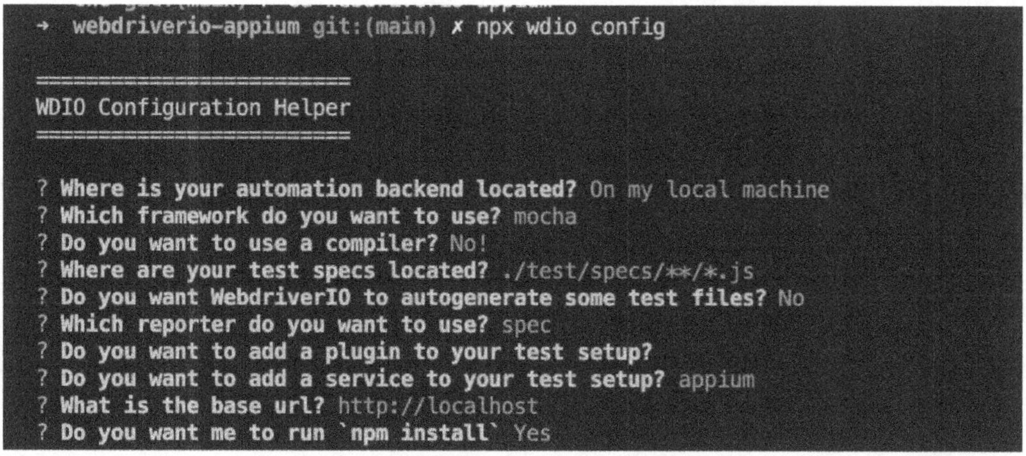

```
→  webdriverio-appium git:(main) ✗ npx wdio config


===============================
WDIO Configuration Helper
===============================


? Where is your automation backend located? On my local machine
? Which framework do you want to use? mocha
? Do you want to use a compiler? No!
? Where are your test specs located? ./test/specs/**/*.js
? Do you want WebdriverIO to autogenerate some test files? No
? Which reporter do you want to use? spec
? Do you want to add a plugin to your test setup?
? Do you want to add a service to your test setup? appium
? What is the base url? http://localhost
? Do you want me to run `npm install` Yes
```

Figure 6.6 – WebdriverIO configuration

Here are the steps

1. In the first step, we will be selecting the location of the automation backend. Since we are not running the tests externally yet, we will be selecting the first option, which is the local machine.

   We will be using the **Mocha** framework for our tests, and that option should be selected in the next step.

2. In the next step, we will be selecting the No option as we will not be using any compiler in our tests.

3. We will go with the default location for our test specs in the next step.

4. We will type No in the next step as we will be starting all our tests from scratch and do not need the autogenerate option.

5. We will stick with the default spec reporter in the next step.

6. Let us skip adding any plugins at this step.

7. In the next step, since we will be running the tests from Appium, we will not be needing any additional drivers. So, let us select the appium option.

8. We can ignore the base URL step as we will not be running a test on the web, and move on to the next step.

9. In the last step, we can select Yes for running the npm install.

The use of the CLI to configure WebdriverIO is productive as we circumvent the need for installing these packages manually. WebdriverIO does a lot of heavy lifting for us to configure the fundamental requisites for our project.

Before proceeding, please make sure to install Appium and its drivers again within the `webdriverio-appium` folder as this is where we will be storing and executing our tests.

In the next section, let us look at additional WebdriverIO configuration for Android.

WebdriverIO Android configuration

As part of the WebdriverIO setup through the CLI, you will notice that a new `wdio.conf.js` file has been created. This is the primary file where we will be making changes to get WebdriverIO working with Appium. Let us now go ahead and look at how it's set up to start making changes.

Quickly browsing through this file shows the customizable port number (`4723`) where the Appium server will spin up. All configurations that we did through the CLI should also be reflected in this file. The important change to be done in this file is in the `capabilities` section. It shows the browser as Chrome by default. Here, we will be specifying the Android settings to connect to the Appium server and run it via WebdriverIO.

Let us now copy our test application within our project by creating a new `app/android` folder structure, as shown in *Figure 6.7*:

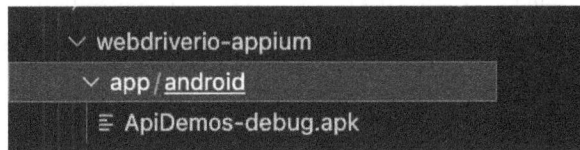

Figure 6.7 – Copying test Android app

We are now ready to update the capabilities section of the config file. Refer to *Figure 6.8* for the values to be used here. We have added the platform name and platform version, which are Android and 13.0, respectively, in this case. Then, we added the device name, which should be the same as the one set up for the emulator in Android Studio. The automation name is the name of the driver used for Android automation.

For the app path, we use the path library to dynamically create a complete path to the test app within our project, as shown in the following screenshot. This library is built-in in Node.js and doesn't need to be installed separately. The path library must be initialized at the beginning of this config file using `const path = require('path')`. This completes the preliminary customization of the config file for the Android application:

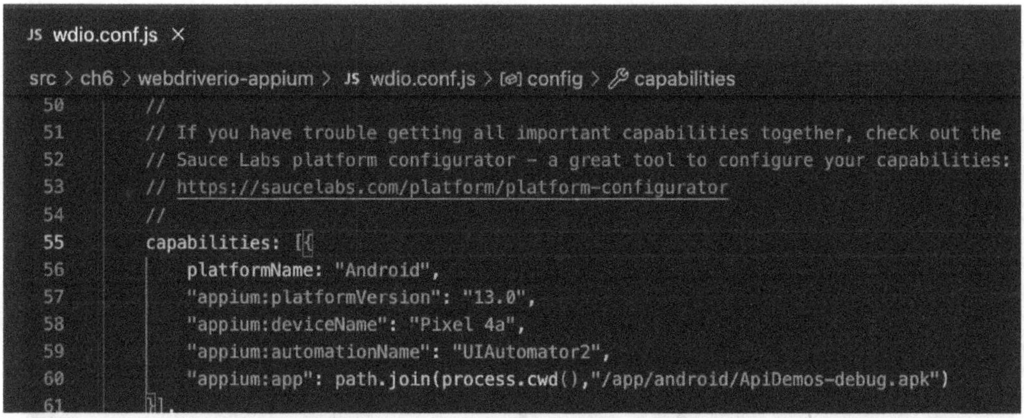

```
JS wdio.conf.js  ×

src > ch6 > webdriverio-appium > JS wdio.conf.js > [@] config > 🔧 capabilities
 50        //
 51        // If you have trouble getting all important capabilities together, check out the
 52        // Sauce Labs platform configurator - a great tool to configure your capabilities:
 53        // https://saucelabs.com/platform/platform-configurator
 54        //
 55        capabilities: [{
 56            platformName: "Android",
 57            "appium:platformVersion": "13.0",
 58            "appium:deviceName": "Pixel 4a",
 59            "appium:automationName": "UIAutomator2",
 60            "appium:app": path.join(process.cwd(),"/app/android/ApiDemos-debug.apk")
 61        }],
```

Figure 6.8 – wdio config: capabilities section

Before we try to run the app with WebdriverIO, let us create a test folder and an empty test file, as shown in *Figure 6.9*. Also, launch the specified emulator from within Android Studio:

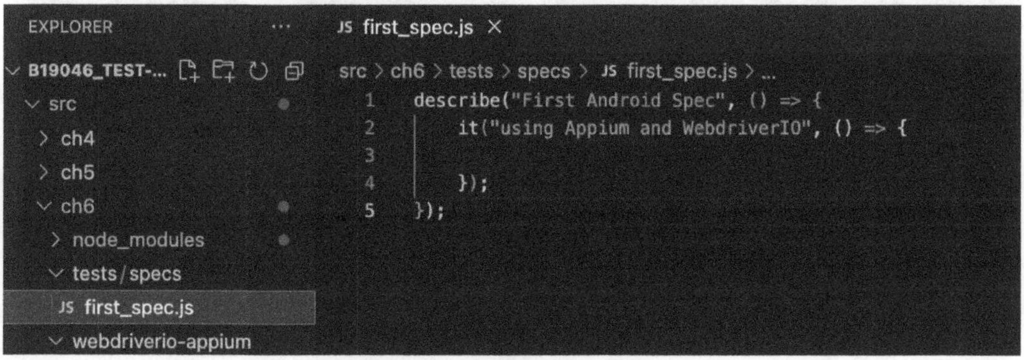

```
EXPLORER                        JS first_spec.js  ×

∨ B19046_TEST-...  ⎘ ⎗ ↻ ⊟    src > ch6 > tests > specs > JS first_spec.js > ...
  ∨ src                     ●    1    describe("First Android Spec", () => {
    > ch4                         2        it("using Appium and WebdriverIO", () => {
    > ch5                         3
    ∨ ch6                    ●    4        });
      > node_modules         ●    5    });
      ∨ tests / specs
        JS first_spec.js
      ∨ webdriverio-appium
```

Figure 6.9 – Test folder and file creation

We are now ready to run this spec. For running this spec, use the `npx wdio` command. This command by default uses the `wdio.config.js` file to spin up the Appium server, load the test app, and execute the test. Results from the empty test can be seen in *Figure 6.10*:

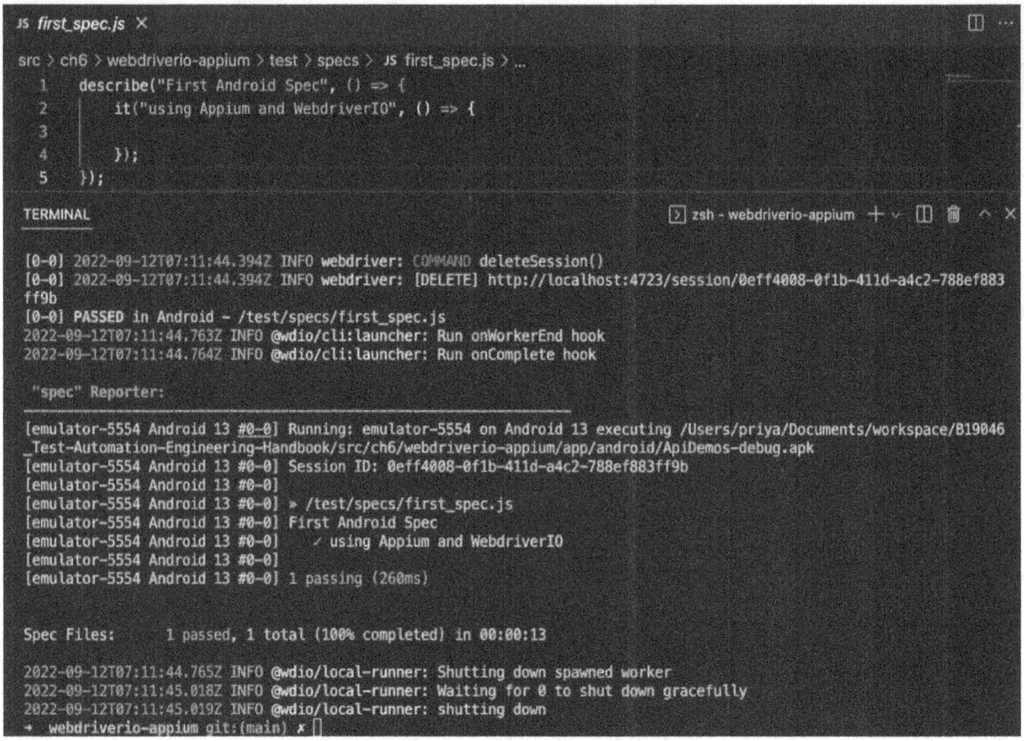

Figure 6.10 – WebdriverIO test run log

This confirms that our installation/setup is complete for Android, and we can go over the manual configuration for the Appium Inspector tool.

Appium Inspector installation and configuration

Let us begin by installing the Appium Inspector, which is a handy tool to inspect mobile elements on the desktop. The latest release can be downloaded from the following link: `https://github.com/appium/appium-inspector/releases`. Opening the application after completing the download and installation would display a dialog, as shown in *Figure 6.11*. In this dialog, when we put in the server information and desired capabilities and hit the **Start Session** button, we will be able to connect the inspector with the emulator. We will look at the details to be filled in here in detail in the subsequent sections:

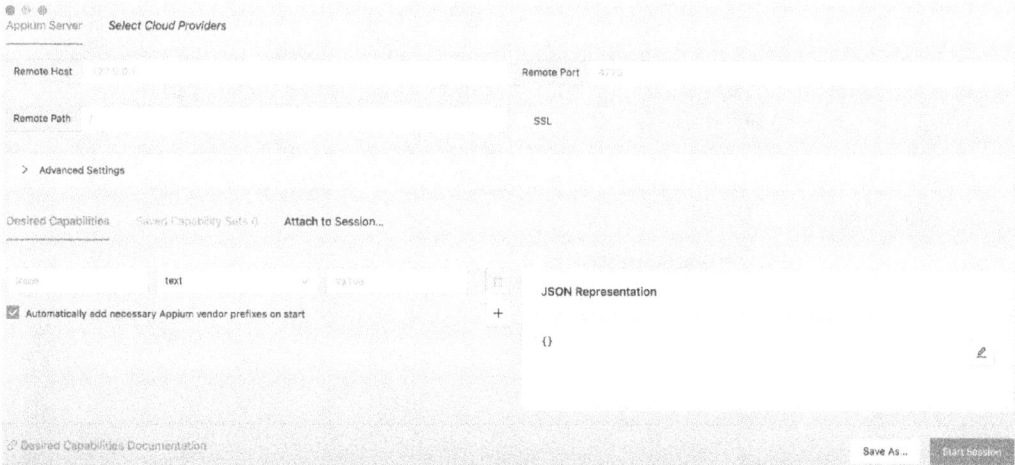

Figure 6.11 – Appium Inspector tool

In the previous section, we used an emulator to run automated tests with WebdriverIO. Now, let us configure another emulator that can be used to find elements on our Android application. Earlier in this chapter, we added a virtual device within Android Studio, which was shown in *Figure 6.6*.

So, we already have two different virtual devices set up—namely, Pixel 4a and Pixel 3a. Now, let us set up Pixel 3a on the Appium Inspector tool. We will be adding capabilities, as illustrated in *Figure 6.12*, under the **Desired Capabilities** section of the tool. Also, remember to update the port number to **4724** since we are already using **4723** for running our WebdriverIO tests:

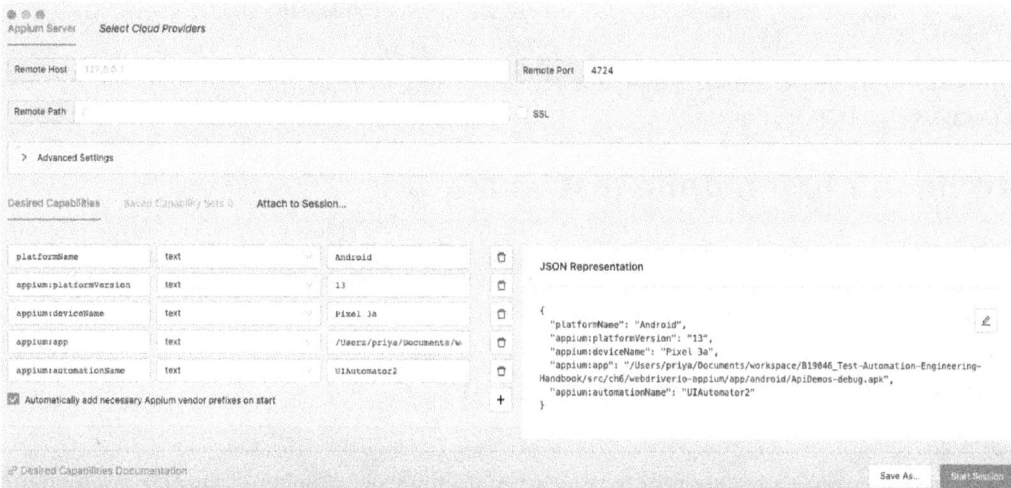

Figure 6.12 – Appium Inspector configuration

Next, bring up the Appium server on port 4724 using the `appium -p 4724` command. Once the server is up, hit the **Start Session** button on the Appium Inspector window to load our application on the Pixel 3a emulator. *Figure 6.13* shows the test application loaded simultaneously in the emulator and Appium Inspector window:

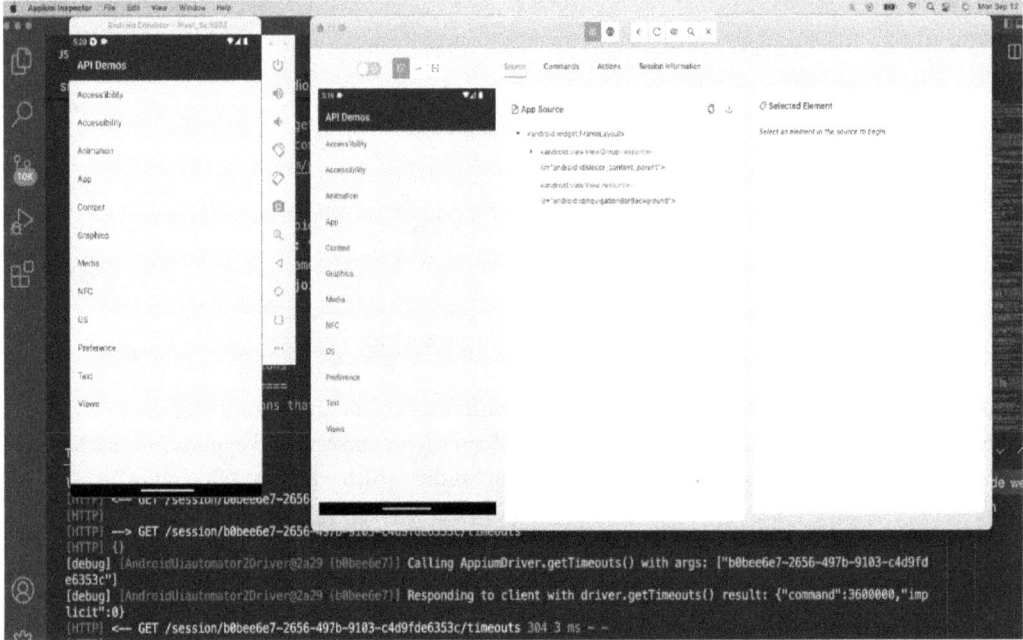

Figure 6.13 – Test application loaded in Appium Inspector and the emulator

In the next section, let's understand how an `async` function works in JavaScript and then write a test to validate mobile elements.

Writing our first mobile test

WebdriverIO tests necessitate the use of JavaScript functions with `async/await` loops. This corresponds to the asynchronous nature of the calls made within the tests. Let us try to understand what these functions are and how to write one.

JavaScript functions with async/await

Let's quickly understand what `async/await` functions are and how they are used in JavaScript. Async stands for asynchronous, and it permits the execution of a function without blocking the flow of the program. It uses promises, which are nothing but values that get fulfilled in the future. For example, let us consider a case where a third-party API is called from within a function as a promise. If the call and its response are successful, then the promise is fulfilled, but if there is a network failure, then it is rejected. We could define specific behavior for each of these cases.

When the `async` keyword is used before a function, it always makes it return a promise. Similarly, the `await` keyword used within the `async` function waits until the promise resolves. `async/await` are essentially the syntactic sugar in JavaScript that let programmers write asynchronous code in a synchronous manner. We will be using `async/await` keywords in the next section to write tests in WebdriverIO.

Figure 6.14 shows an example of a simple function with a promise using `async/await` keywords. A promise named `simple_promise` is being created, and it is set to be resolved after a timeout of 5 seconds. The result of `simple_promise` is stored in the `result` variable, which is then printed onto the console:

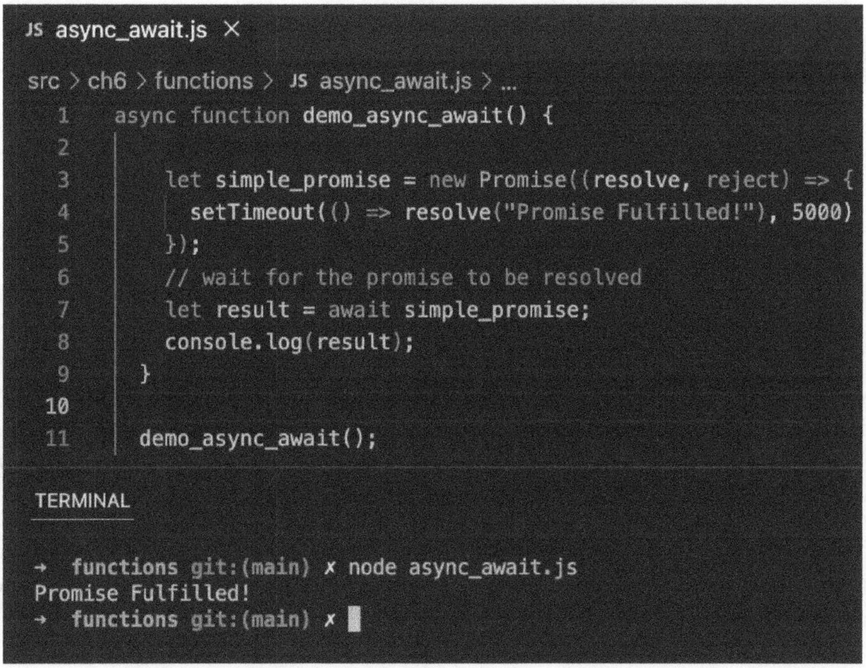

```
JS async_await.js  ✕

src > ch6 > functions > JS async_await.js > ...
    1    async function demo_async_await() {
    2
    3        let simple_promise = new Promise((resolve, reject) => {
    4            setTimeout(() => resolve("Promise Fulfilled!"), 5000)
    5        });
    6        // wait for the promise to be resolved
    7        let result = await simple_promise;
    8        console.log(result);
    9    }
   10
   11    demo_async_await();

TERMINAL

→  functions git:(main) ✗ node async_await.js
Promise Fulfilled!
→  functions git:(main) ✗ █
```

Figure 6.14 – JavaScript function with async/await

That was a very quick introduction to promises and JavaScript functions with `async/await` loops. There is definitely more to it, and these functions can get extremely complicated easily. We will just be using a simple `async/await` loop in our WebdriverIO test in the next section. Let us now get to writing a test for a mobile application.

First Appium/WebdriverIO test

We are finally ready to write our WebdriverIO test and run it in the Android emulator using Appium. Let us use the spec `first_spec.js` file to identify an element using its accessibility ID and perform an assertion on it. The advantage of using this selector is that it is cross-platform compatible. This avoids the need to have a platform-specific locator strategy.

Another advantage of using this selector is that it does not change with localization. A straightforward way to do this would be by loading the Android app in the Appium Inspector tool, as shown previously in *Figure 6.13*, and clicking on the element to be identified from the leftmost pane. This displays the complete set of properties of that element on the rightmost pane where its accessibility ID can be found. *Figure 6.15* demonstrates this selection where the **Animation** option is selected and its properties are shown:

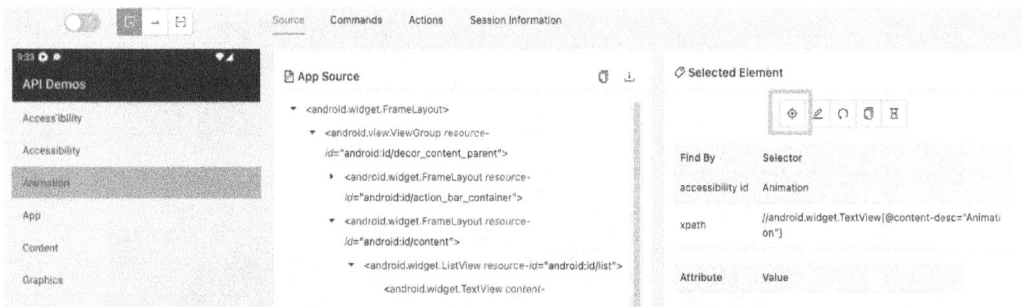

Figure 6.15 – Inspecting an element via Appium Inspector

Let us now write a simple test to click on the **Animation** option and verify that the **Bouncing Balls** option is displayed on the subsequent screen. The user can navigate to the menu within **Animation** by using the highlighted **Tap** icon in *Figure 6.15*. We begin the test by capturing the **Animation** option using its accessibility ID. The tilde symbol (~) is used for that.

Then, we use the click () function to interact with the captured element. In the next menu, we capture the **Bouncing Balls** element and verify its presence with the toBeExisting() function. The test can be triggered using the npx wdio command:

```
describe ("First Android Spec", () => {
it ("to find element by accessibility id", async () => {
const animationOption = await $("~Animation");
await animationOption.click();
const bouncingBalls = await $("~Bouncing Balls");
await expect(bouncingBalls).toBeExisting();
});
});
```

Figure 6.16 shows the results of the test execution, with the Appium Inspector showing the selected menu option:

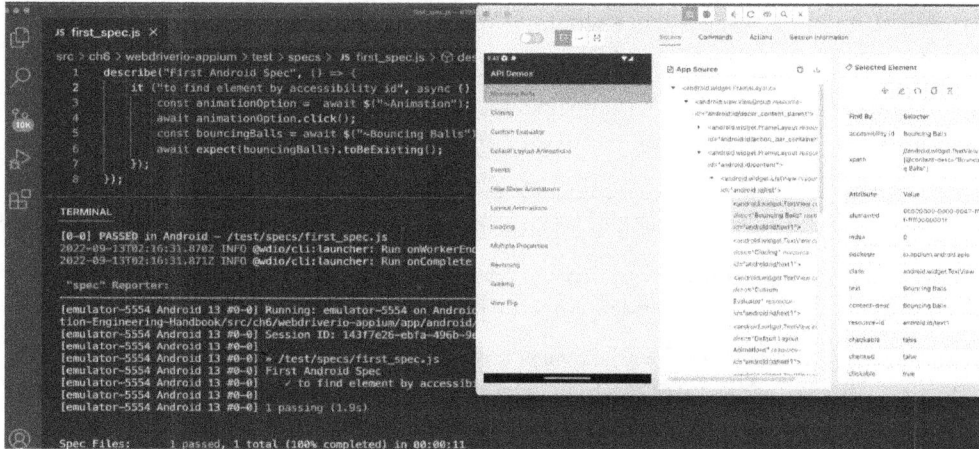

Figure 6.16 – Appium test execution

Other selector options to explore are `class`, `xpath`, and `elementId`. When using these selectors, users should be wary of their uniqueness and how concise they are. This completes our section for Android mobile automation. We went through the installation and setup process and got to the point where we are able to write our own tests. For further exploration of the various features that Appium offers, refer to the documentation at `https://appium.io/docs/en/about-appium/api/`.

In the next section, let us survey a few key considerations for mobile automation.

Key considerations for mobile automation

Mobile automation is different from web automation in many ways, and it gets complicated with the various versions of the operating systems and the use of emulators. Let us now look at some common areas of complexity to look for when working on mobile test automation.

Areas of complexity

Mobile automation poses certain challenges that are common across platforms. Let us examine them one by one now:

- One of the major challenges is that a mobile application can be run on different devices from distinct manufacturers on multiple versions of operating systems. A solid automated test strategy addressing these compatibilities is a must-have.

- Usability testing to assure a consistent user experience across these platforms must be planned and executed within the testing cycle rather than being an afterthought.

- Testing for various geographical regions and cultures must be performed to verify the accuracy and relevance of the mobile application.

- Every platform has its own native actions, and test automation should account for them and design tests to verify them.

- Every mobile application must be thoroughly security-tested for possible threat and vulnerability scenarios.

- Testing for a range of internet speeds and bandwidths should also be performed to verify the stability of the mobile application in urban as well as remote areas.

- Mobile devices come with diverse screen resolutions, and it is the role of test automation to validate that the application renders correctly in all of them.

- Test automation should consider the type of mobile application being tested. The test approach for a native application that is designed to work only on a specific device should be different from a hybrid application. For example, **Progressive Web Apps (PWAs)** are expected to work on any device and therefore involve a more diverse testing effort.

Let us next review some iOS-specific factors that influence test automation.

iOS-specific considerations

Creating automated tests for the iOS platform needs distinct installations and further setup. Let us list here the various pieces that work together to write a functional iOS test:

- iOS has a separate development environment called **Xcode**, and this is used to spin up the emulator's specific iOS platform. This can be downloaded from the AppStore.

- **Xcode Command Line Tools** comes with many useful tools such as Apple LLVM, compiler, linker, and Make for easy execution of software through the command line:

  - **Carthage**, which is a dependency manager for iOS and macOS

  - **Appium doctor**, which we have already used to verify the setup and the presence of all the required dependencies

  - **XCUITest, which** is the primary Appium driver to test iOS applications

Here is an example of the capabilities section of the wdio.config.js file for iOS:

```
platformName: "iOS",
"appium:platformVersion": "15.0",
"appium:deviceName": "iPhone 13",
"appium:automationName": "XCUITest",
"appium:app": path.join(process.cwd(),"/app/iOS/Test-iOS.app")
```

Let us look at the pros and cons of using emulators for automated mobile testing in the next section.

Real devices versus emulators

The chief aspect of mobile automated testing lies in choosing between emulators and physical devices. You might say that testing against a physical device is more realistic, but it comes at a cost. At the same time, testing with emulators has certain drawbacks. Let us explore the differences in detail here.

An emulator is a software program that simulates the hardware and operation of the mobile device. Its main goal is to provide the user with an experience of using a real device. It is usually platform-specific and is customizable to use different versions of the hardware and software. One of the key advantages of using emulators for test automation is their cost. Their easy availability also makes them a viable option for testing. Unlike real devices, they also come with certain features to aid in testing. Their main downside is in replicating certain performance issues such as battery life and network connectivity. The test results in these areas may be unreliable. It also gets increasingly complicated to test the device behavior when multitasking. This is where using real devices come in very handy.

Testing on real devices includes running automated tests in the same way the end users would use handsets. The results in this case are highly reliable and can be effectively used to validate battery and other performance issues. Using real devices comes with a high cost, though. Every model with every upgrade must be bought and maintained in terms of both hardware and software. Cross-platform testing becomes impossible with real devices as the OS must be updated for every change. A good mobile test automation strategy should include a combination of both real devices and emulators. This gives the right balance for testing both functional and performance aspects of the mobile device with its dependencies on the platform. Let us now look at how **cloud service providers** (**CSPs**) assist in mobile test automation.

Choosing CSPs

Cloud-based mobile application testing provides the option to users to link with multiple mobile devices at the same time. It is a key aspect of today's mobile automation strategy. It enables us to scale testing with ease while testing features across multiple geographic locations. Organizations can choose to use a private cloud for their testing, which drastically reduces security concerns. Cloud-based providers give you options to combine various versions of hardware and software for testing. Usually, the number of such combinations available to the users is in the scale of thousands. This allows us to test and debug issues much quicker. Testing teams can focus purely on the functional and performance aspects while leaving the device/emulator maintenance to the **service providers** (**SPs**). This also avoids the cost of setting up an on-premise testing lab.

Some notable cloud providers are as follows:

- Sauce Labs
- BrowserStack
- LambdaTest

- Kobiton

- AWS Device Farm

- Firebase Test Lab

These providers have a predefined subscription and offer several advanced built-in features to suit testing needs. They come up with a default test pipeline setup that can also be run in parallel to save time. An important consideration while choosing a cloud-based provider is to review their quality and security standards before entering into a contract with them. The cost of moving out of a cloud-based mobile platform rises the longer you are with them. Let us review some ways to optimize our mobile test framework.

Optimizing our mobile automation framework

Let us now look at a couple of ways to optimize our framework. The ability to switch between Android and iOS tests without manually changing the configuration each time is a useful addition to our test framework. This can be easily achieved in WebdriverIO by separating out the `wdio.conf.js` files for these platforms. *Figure 6.17* shows the setup of a `config` folder with respective config files:

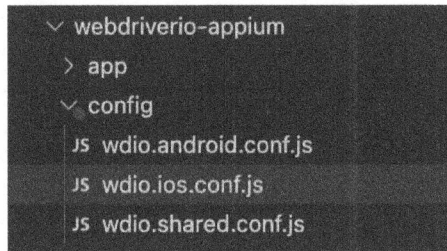

Figure 6.17 – WebdriverIO config file setup

Each of the config files for the platforms will contain platform-specific information, as shown in *Figures 6.18* and *6.19*. The corresponding sections should be removed from the shared config file:

```
JS wdio.shared.conf.js        JS wdio.android.conf.js  ×     JS wdio.ios.conf.js

src > ch6 > webdriverio-appium > config > JS wdio.android.conf.js > ...
  1     const { config } = require('./wdio.shared.conf')
  2     const path = require('path');
  3     config.port = 4723
  4     config.specs = [
  5         '<<folder where the Android tests live>>'
  6     ]
  7     config.capabilities = [
  8         {
  9             platformName: "Android",
 10             "appium:platformVersion": "13.0",
 11             "appium:deviceName": "Pixel 4a",
 12             "appium:automationName": "UIAutomator2",
 13             "appium:app": path.join(process.cwd(),"/app/android/ApiDemos-debug.apk")
 14         }
 15     ]
 16     exports.config = config
```

Figure 6.18 – WebdriverIO Android config file

The Android config file holds the path where the Android tests live and their platform-specific capabilities:

```
JS wdio.shared.conf.js        JS wdio.android.conf.js        JS wdio.ios.conf.js  ×

src > ch6 > webdriverio-appium > config > JS wdio.ios.conf.js > ...
  1     const { config } = require('./wdio.shared.conf')
  2     const path = require('path');
  3     config.port = 4723
  4     config.specs = [
  5         '<<folder where the iOS tests live>>'
  6     ]
  7     config.capabilities = [
  8         {
  9             platformName: "iOS",
 10             "appium:platformVersion": "15.0",
 11             "appium:deviceName": "iPhone 13",
 12             "appium:automationName": "XCUITest",
 13             "appium:app": path.join(process.cwd(),"/app/iOS/Test-iOS.app")
 14         }
 15     ]
 16     exports.config = config
```

Figure 6.19 – WebdriverIO iOS config file

The iOS config file similarly holds iOS-specific information. The port value remains the same in these files, but it may change when executed using a cloud-based provider for these platforms. Another change would be the way we execute these tests. Now that we have distinct config files, we will run them using the `npx wdio config/wdio.android.conf.js`command.

Another quick way to optimize our tests is by using hooks. Hooks are essentially reusable code that can be run multiple times but defined only once. The commonly used hooks in WebdriverIO are `before`, `beforeEach`, `after`, and `afterEach`. The `before` hook runs before the first test in the block, and the `after` hook runs after the last test in the block. `beforeEach` and `afterEach` hooks run before and after each test in the block. Hooks can help reduce test dependency, which in turn helps reduce code duplication. All prerequisites for a given block/test can be grouped into `before` and `beforeEach` blocks. Similarly, all cleanup and teardown steps can be organized into `after` and `afterEach` blocks.

This brings us to the end of this chapter. Let us quickly summarize what we have learned in this chapter.

Summary

In this chapter, we covered various aspects of mobile test automation. We started with understanding what Appium and WebdriverIO are. Then, we dived into their installation and setup in detail. We learned how to set up and execute tests in WebdriverIO using Appium. While doing this, we also absorbed how to use the Appium Inspector tool to find elements and used it in our spec.

Then, we moved to review certain key considerations for mobile test automation. In that section, we surveyed the areas of complexity, understood the difference between using real devices and emulators, and investigated the advantages of using CSPs. Finally, we picked up a couple of ways to optimize our test framework.

In the next chapter, we will be taking up the exciting topic of test automation in the API world. We will be using the **Postman** tool to achieve that. We will also learn about using Docker containers to execute our API tests.

Questions

1. What is Appium doctor used for?
2. What are the disadvantages of using an emulator for mobile testing?
3. Name some common challenges with mobile test automation.
4. What are the benefits of using CSPs for mobile test automation?
5. How do hooks help in the test framework?

7

Test Automation for APIs

An **application programming interface** (**API**) helps two systems interact with each other over the internet by implementing a set of definitions and protocols. APIs help tremendously to simplify the design of software systems and assist in opening up access to the system while maintaining control and security over calls being made. APIs primarily act as intermediaries between various parts of the application as well as with external integrations.

There are three main kinds of API architectures—namely, **REST**, **RPC**, and **SOAP**. Each one of them has its pros and cons. Diving deep into the types of APIs and their architectures is out of the scope of this book. Instead, we will be learning about testing **Representational State Transfer** (**REST**), which has been the most popular architecture for constructing APIs.

It is crucial to maintain the quality standards of APIs as they are often the point of entry for various external application calls. Therefore, API testing becomes a fundamental part of the software development life cycle. There are various testing tools available in the market for testing APIs, some of the notable ones being **Postman**, **SoapUI**, **Apigee**, **Assertible**, and so on. Each of these tools comes with various capabilities, and it is vital to choose the right tool that suits the needs of your organization. In this chapter, we will be looking at API testing with Postman. Postman comes with plenty of built-in tools, and most of them are available for free. These are the topics that we will be covering in this chapter:

- Getting started with Postman
- Sending GET and POST requests
- Writing automated API tests
- Key considerations for API automation

Technical requirements

We will continue using Node.js along with JavaScript in this chapter. We will also download the Postman application (version 9.3.15) and the Newman command-line tool. We will need Docker installed locally to run our Postman collections on a Docker container. Docker installation instructions can be found at https://www.docker.com/.

Getting started with Postman

Before we dive into the Postman tool, it is worthwhile to understand the basics of REST APIs and what goes into testing them.

Basics of REST API testing

A REST API is stateless in nature and is built on client-server architecture. The client must provide all the data such as headers, authorization, and so on for every request since the server cannot store any information due to the stateless nature of these APIs. The communication usually happens over HTTP using JSON with support for **Create, Read, Update, Delete** (**CRUD**) methods. Some of the most commonly used REST API methods are the following:

- GET: Retrieve information about a REST API resource
- POST: Create a REST API resource
- PUT: Update a REST API resource
- DELETE: Delete a REST resource or a related component
- OPTIONS: List the supported operations in a web service
- HEAD: Returns only HTTP header and no body

Syntax to call a RESTful API goes like this: http://{host}:{port}/{initialPath}.

Here, the following applies:

- The {host} placeholder indicates the host or domain name
- The {port} placeholder indicates the TCP port number
- The {initialPath} placeholder indicates any initial path that is part of the URI for a given deployment

The primary focus of API testing is to validate the business logic of APIs, while other types of testing include performance and security. It involves the collection of test data to be sent in as JSON requests to the API and validating the response for accuracy. Some of the items to look for in the response are status, syntax, schema conformance, and functional correctness. API testing is much quicker and provides the benefit of testing very early in the development life cycle. Let us now get started with downloading and setting up Postman.

Downloading the Postman application

Postman is a popular and rapidly evolving tool that helps create, test, and document APIs. It is mainly an open source tool but has certain advanced features bundled under a **Pro** version. It supports **Java**, **JavaScript**, and **Groovy** programming languages and is quite easy to learn. It is primarily used for performing exploratory and automated API tests. Let us now download the Postman application.

To get started with the download, click on the button corresponding to your machine configuration on Postman's downloads page (`https://www.postman.com/downloads/`). Open the downloaded ZIP file, and you should have the Postman app. Clicking on the icon opens the application where the user is provided with an option to sign in or create a new account. It is strongly recommended to have an account to be able to store your requests and configurations for future use. Users should see a window like the one shown in *Figure 7.1* upon signing in to the account:

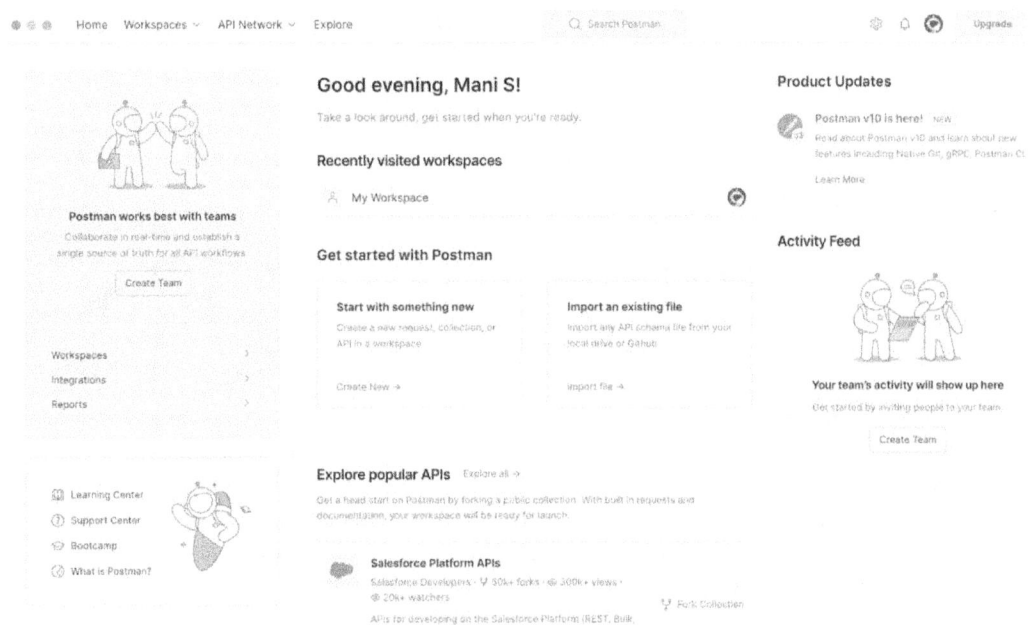

Figure 7.1 – Postman application

In the next section, let us look at how to work with workspaces in Postman.

Creating and managing workspaces

Workspaces are at the center of all the action that happens in Postman. It is one of the most important features that allows collaboration with your team and helps assemble your API collections. A workspace primarily aids in organizing the project according to the individual's needs. Workspaces can be shared with your team members or worked on personally as well. Let us now look at how to create a new workspace in Postman. From the home page of the application shown in *Figure 7.1*, click on the **Create New** link to open a modal showing all the available options. Select the **workspace** option and provide a name and a brief summary for this workspace, as shown in *Figure 7.2*. Users have the option to select the type of access for this workspace:

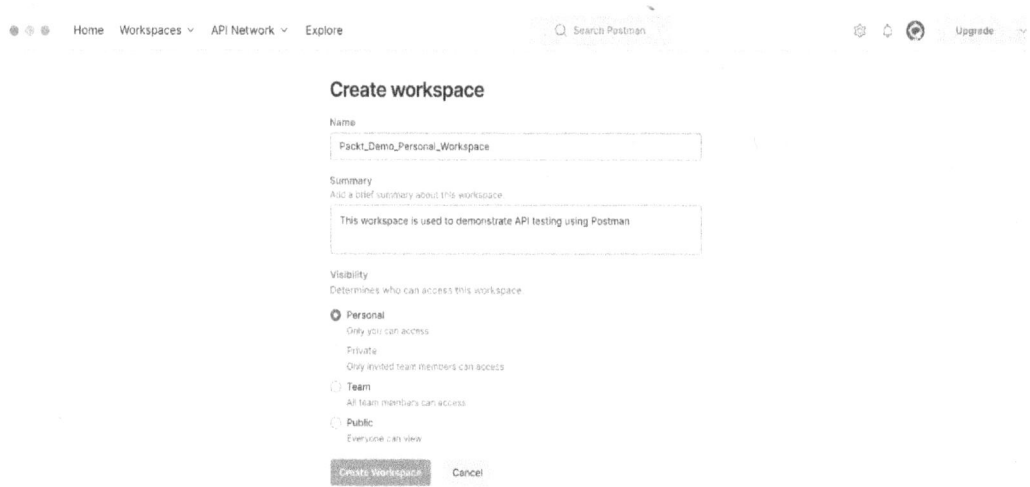

Figure 7.2 – Creating a workspace

We are creating a **Personal** workspace for this example. **Private** and **Team** workspaces allow adding team members to the workspace by inviting them via email or by sharing the workspace link, whereas a **Public** workspace can be accessed by anyone working with the Postman tool. The changes done by the team members are synchronized in real time across the user accounts accessing the workspace. *Figure 7.3* shows the home page with multiple workspaces. Users can access the contents of these workspaces on any device they log in to with their account credentials. Thus, workspaces promote effective collaboration across the team:

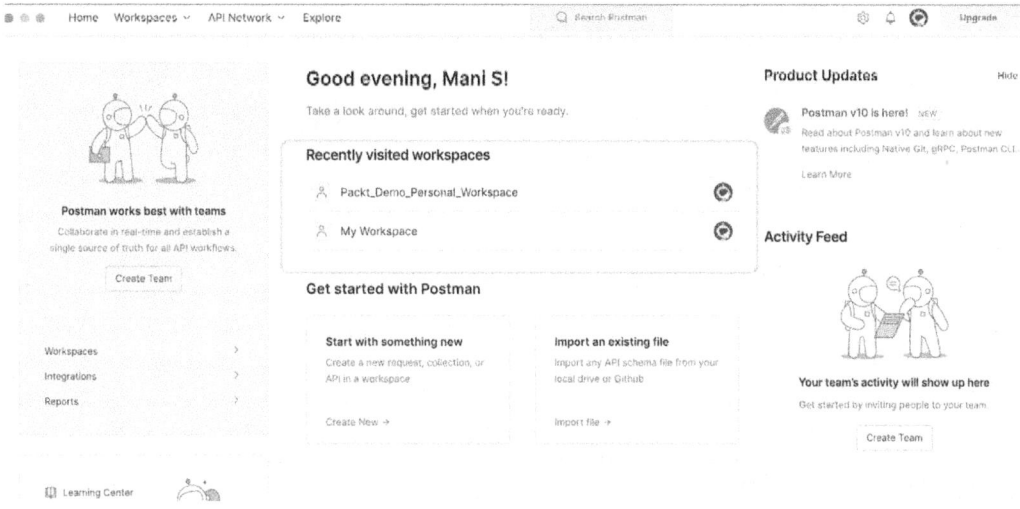

Figure 7.3 – Home page with multiple workspaces

Let us now send our first GET and POST requests using Postman.

Sending GET and POST requests

For creating requests, the user must be within a workspace and click on the + button next to the **Overview** tab. This is the simplest way to create a new API request. A new tab with various request details opens, as shown in *Figure 7.4*. This is commonly referred to as a request dashboard in Postman:

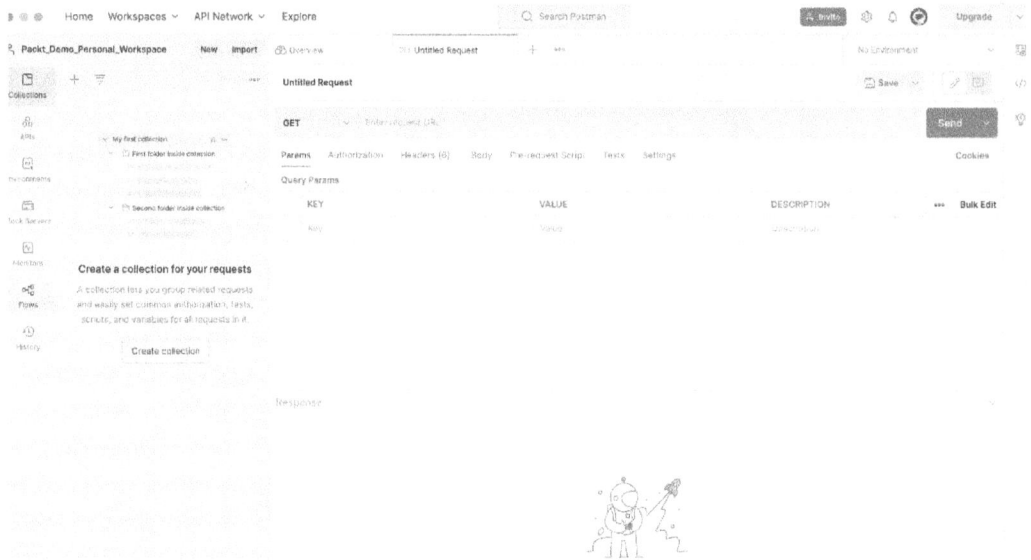

Figure 7.4 – New API request

Now, let us see how to make a GET API request.

Making a GET API request

We will be using the `https://www.boredapi.com/api/activity` API to get an understanding of GET requests. Bored is a free API that returns some random activities to do when bored. Postman makes it easy to get a simple GET API request without any authorization working. Just paste the API URL in the URL section of the request window and hit the **Send** button next to it. Every request to the server must be made with a URL to fetch the required response. *Figure 7.5* shows the GET request with the response. Here, the **Status** section of the response says **200 OK**, which means that the server responded to the request without any errors. The server returns the response in a JSON format, which can be validated for accuracy based on the business logic. In our case, we see an activity being returned with a bunch of other information:

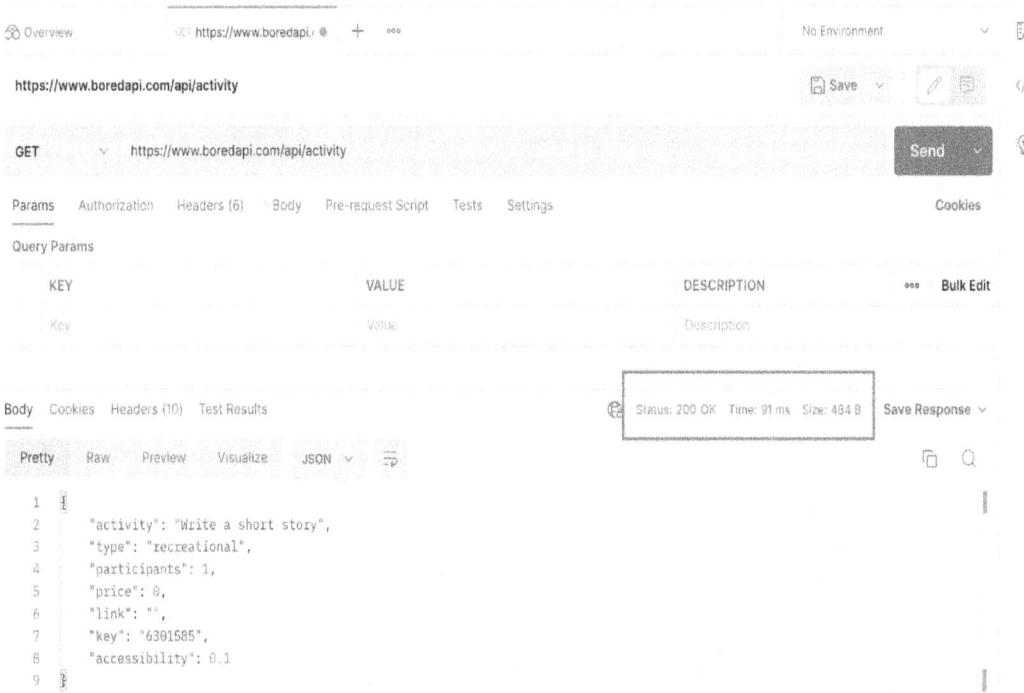

Figure 7.5 – GET API request

In most cases, the API will have certain authorization to be added for the request to work. Postman supports a wide variety of authorization mechanisms that can be accessed via the **Authorization** tab, as shown in *Figure 7.6*:

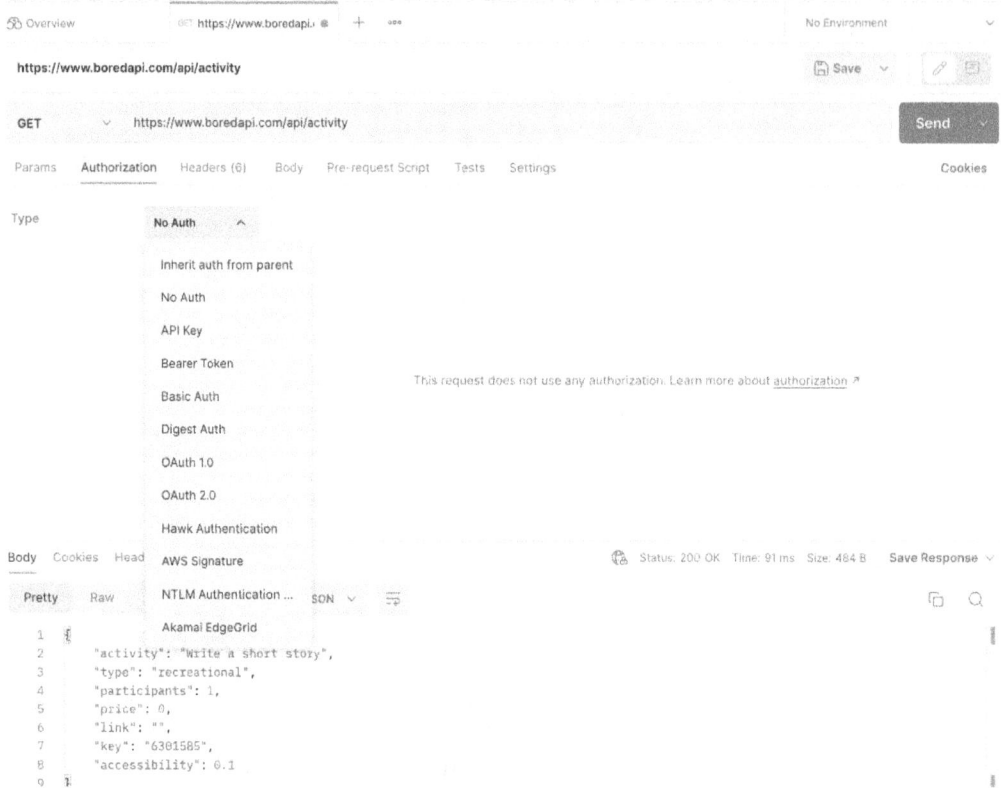

Figure 7.6 – Request authorization support in Postman

Postman identifies the applicable headers for a given API call, but in cases where there is specific metadata that must be sent as part of the header, this can be done using the **Headers** tab. Postman automatically identifies any query parameters that are added as part of the URL. For the case in *Figure 7.7*, `https://api.agify.io?name=packt`, a name is sent as a query parameter as part of the URL. Postman creates a parameter in the **Params** tab, and this can be modified to feed the request with different test data. New parameters can be manually added here as well:

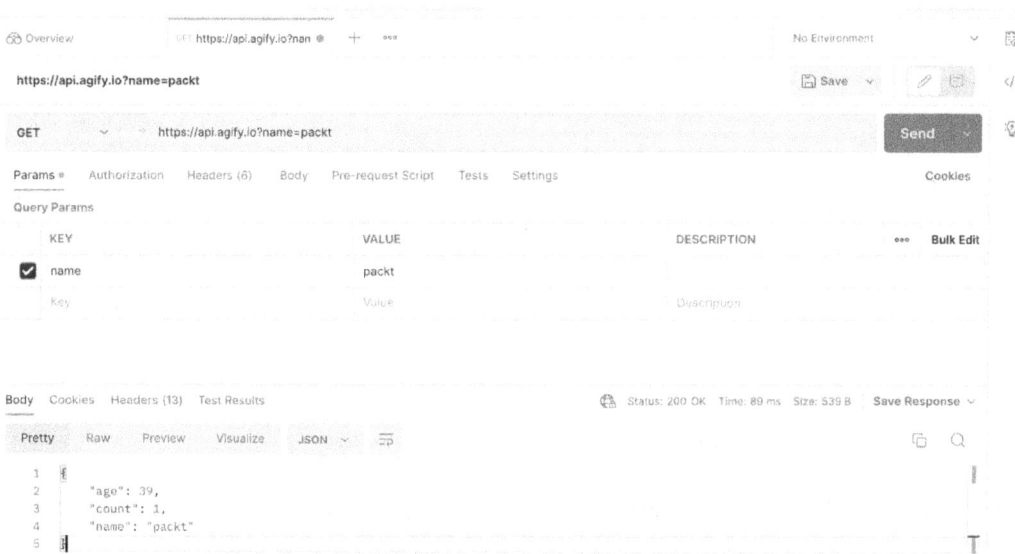

Figure 7.7 – API request with a query parameter

Users are encouraged to check out the various other features available within the request and response windows. Let us now learn to make a POST API request.

Making a POST API request

A POST request creates a new resource on the server and requires content to be sent in the body of the request. Postman supports different body types for a POST call, and in this section, we will review how to make a POST call. Create a new API request and click on the dropdown to the left of the URL section to select a POST request type. We will be using GitHub's *create a new repository* API call (https://api.github.com/user/repos) for our example here. GitHub provides a lot of public APIs, but it requires the generation of an access token. This token can be generated in the personal access tokens section of your GitHub profile. Please remember to copy and save this secure token for future use in Postman. As shown in *Figure 7.8*, use this as a bearer token in the **Authorization** tab of the new POST request:

Figure 7.8 – POST API request authorization

Now, moving on to the **Body** section of the request, this API requires a name as a mandatory key for the new repository being created. An example of this request can be found at this link: `https://docs.github.com/en/rest/repos/repos#create-a-repository-for-the-authenticated-user`. We will be using the raw body type with JSON from the dropdown, as shown in *Figure 7.9*. Postman supports a wide variety of request body formats, and the one supported by the API being tested should be used. On hitting the **Send** button, we complete the POST call to the GitHub server to create a new repository with the name `Packt-test-api-repo`. *Figure 7.9* shows the response status of **201 Created** with all the metadata in the response body section. Users may notice that the status code is different from the GET call as 201 indicates that in addition to the call being successful, a new resource was created by the server:

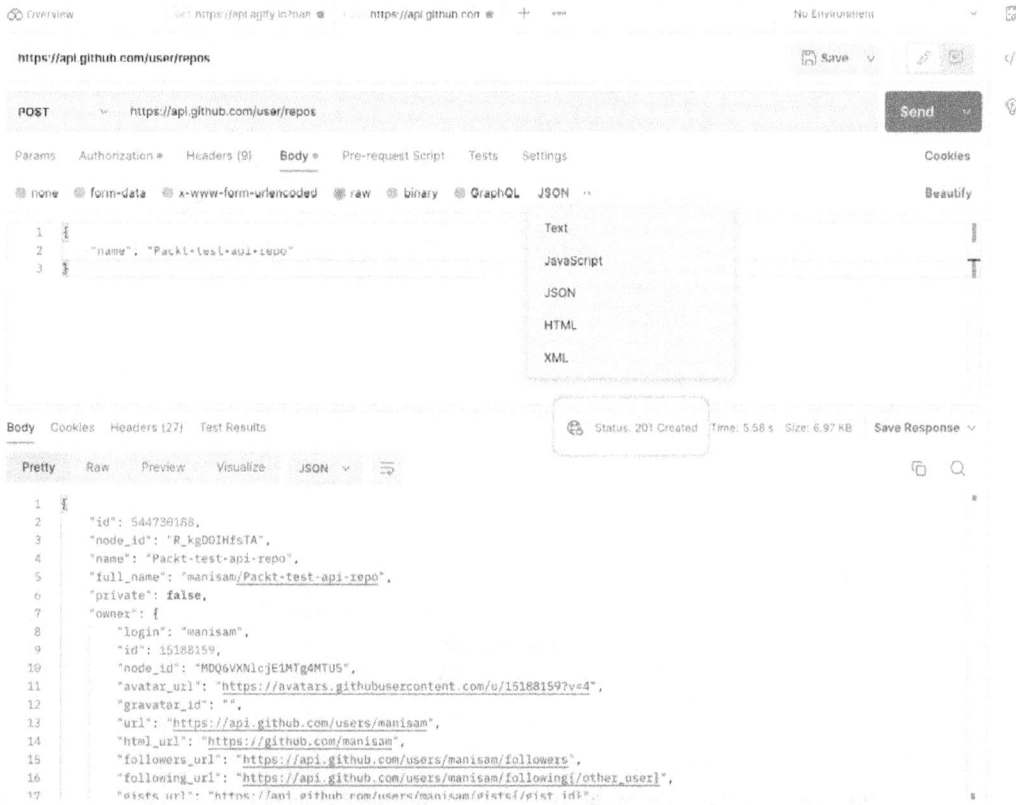

Figure 7.9 – POST API request body type

This completes our section on creating a new resource using a POST call. In the next section, let's learn about collections and how they help structure API requests in Postman.

Organizing API requests using collections

Postman provides a way to group the API requests using collections. It helps organize a workspace by breaking it down, and a workspace can also be sorted into multiple collections. Apart from this, collections can also be published as documentation as well as run together in an automated fashion. In this section, let us review how to create a collection and add requests to it. Collections can be created in multiple ways within a workspace, and a simple way is to use the + button next to the **Collections** icon in the left pane of the **Workspaces** window. A name must be provided for the collection, and both existing and new requests can be saved to this collection. *Figure 7.10* shows a new collection holding the two API requests we have created so far in the previous sections:

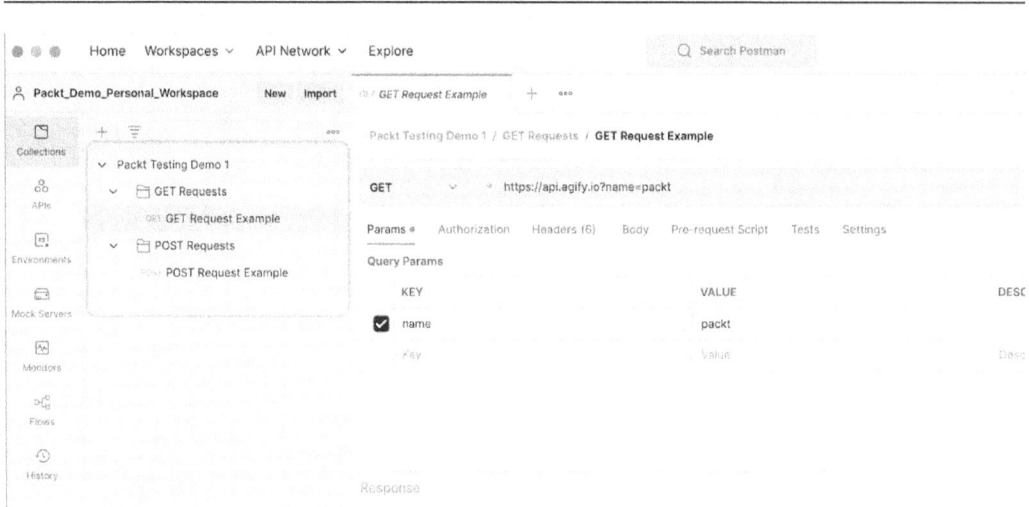

Figure 7.10 – Collections

Collections can thus help organize API requests in a meaningful way under a given workspace. This helps immensely when there are a high number of requests, which is usually the case when testing enterprise applications. Various other actions can be performed on collections, such as **Export**, **Monitor**, **Mock**, and so on. Let us now look at one more feature of a collection that promotes collaboration within the team.

Postman allows users to create a fork of a collection and merge it after making some modifications to the forked collection. These changes can be shared with other members, just like how Git pull requests work. This can be done via the **Collections** drop-down menu, as shown in *Figure 7.11*:

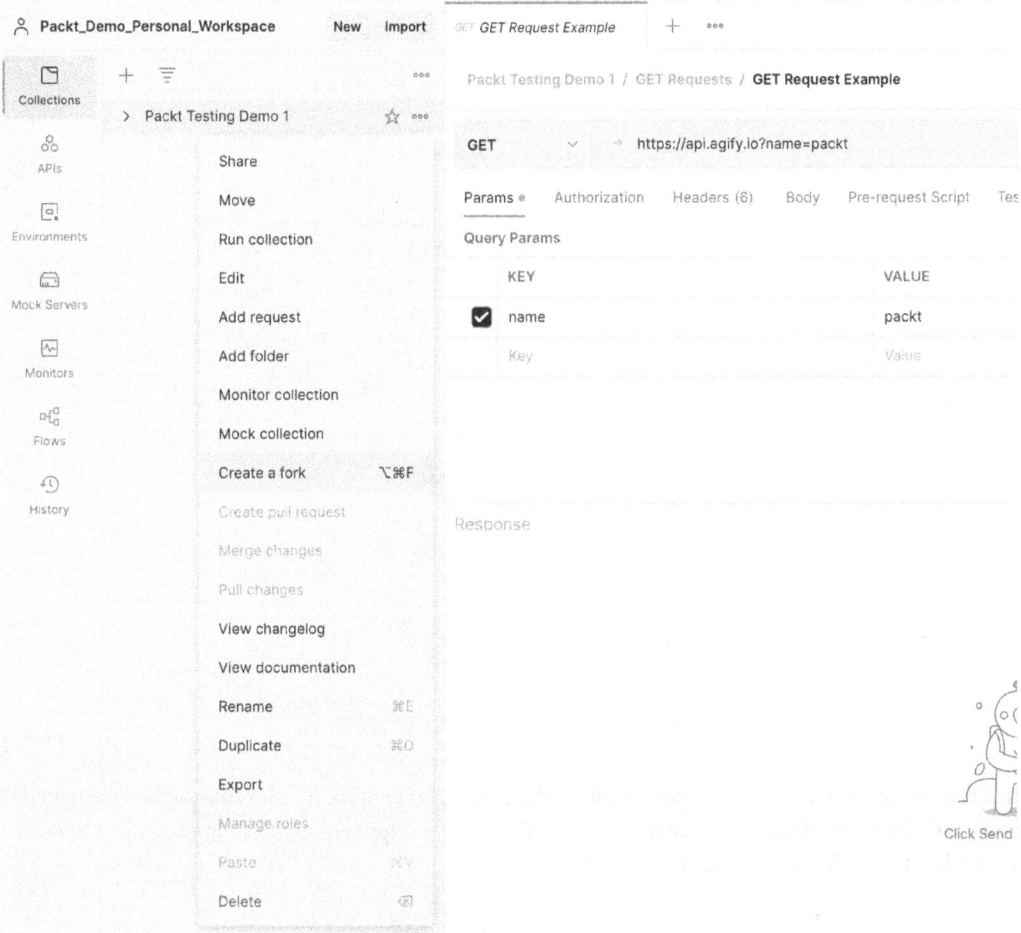

Figure 7.11 – Collections drop-down menu

On clicking the **Create a fork** option, the user will be required to enter a label for the fork and to which workspace this collection is to be forked, as shown in *Figure 7.12*:

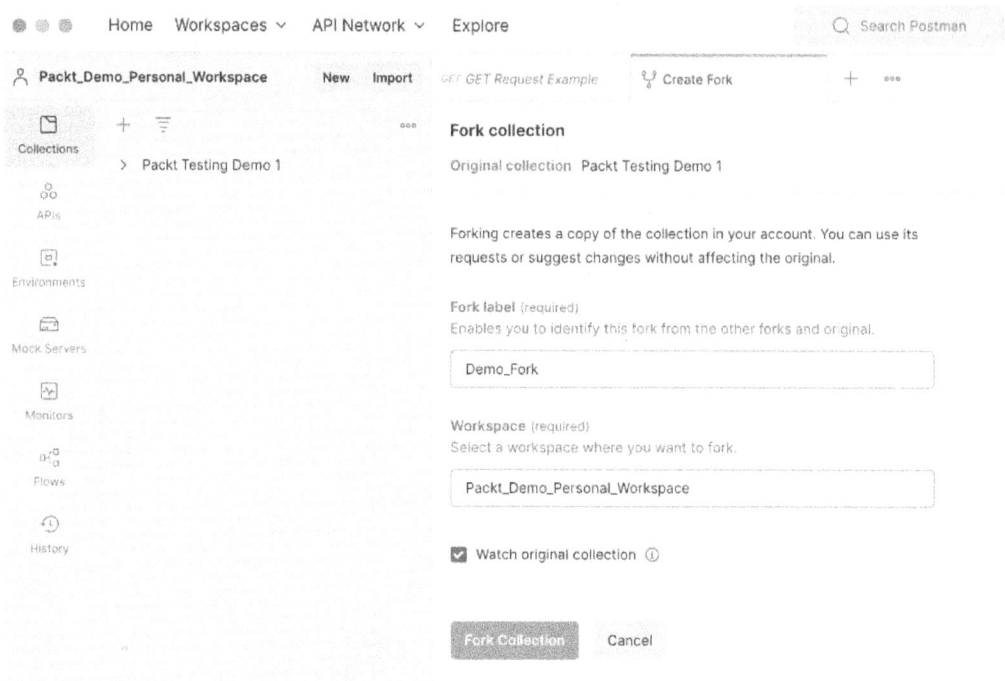

Figure 7.12 – Forking a collection

Once the required changes have been made to the forked collection, a pull request can be created using the **Collections** drop-down menu. *Figure 7.13* shows a snippet of a pull request where the user can provide a title, description, and select reviewers:

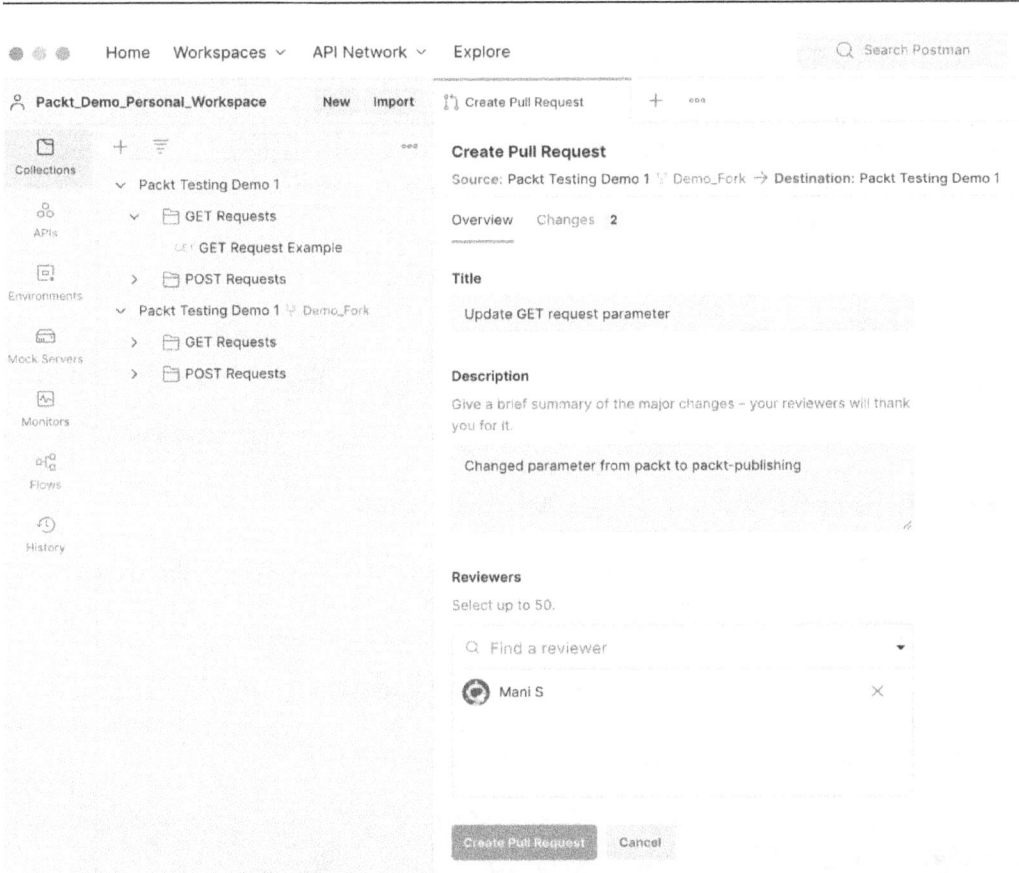

Figure 7.13 – Creating a pull request

The changes can be reviewed and merged using the merge option within the pull request or through the **Collections** drop-down menu. This is a neat way to keep track of changes to your API requests and nurture collaboration within the team while making these changes.

So far, we have looked at the basics of the Postman tool and how to make requests manually. This works well for testing new API features but falls short when it comes to regression testing. In the next section, let's learn to write automated tests to validate API responses.

Writing automated API tests

Postman allows us to add tests that run automatically after an API response is returned from the server. This can be done through the **Tests** tab in the Postman request dashboard. We will be using the following three GitHub API requests in this section to help us set up and understand automated API response validation:

- Create a new repository
- Get a repository by name
- Delete a repository by name

In the next section, let us review how to use snippets to speed up our test automation process.

Using snippets for asserting an API response

Postman comes with pre-defined JavaScript test scripts in the form of code snippets that can directly be used in our tests. Let us start by adding snippets for some of the basic checks performed on an API response. Snippets are shown on the right pane next to the various tabs on the request dashboard.

Every API test requires the validation of the status of an API response, and Postman provides a readymade code snippet for this. On selecting the `Status code: Code is 200` snippet, the following code is populated onto the **Tests** tab, as shown in *Figure 7.14*:

```
pm.test("Status code is 200", function () {
pm.response.to.have.status(200);
});
```

pm represents a Postman object, and it contains all information pertaining to the request and response body. The pm object comes with a lot of properties and methods built in. Here, we use the `test` method, which accepts a description and a function within which an assertion can be defined. These assertions are defined using the `chai` library, and readers can refer to their documentation at `https://www.chaijs.com/api/bdd/` to get familiarized with more assertions.

Let us now add the `Status code: Code name has string` snippet, which adds the following code snippet. In the case of a `GET` request, this string is `OK`, and for the `POST` request, it is `Created`, as we have seen before:

```
pm.test("Status code name has string", function () {
pm.response.to.have.status("Created");
});
```

Postman provides a snippet to validate the response time of the API. On adding the `Response time is less than 200ms` snippet, we see the following code added. Note that we use the `expect` assertion here, which operates on the `responseTime` property and checks its value range:

```
pm.test("Response time is less than 200ms", function () {
pm.expect(pm.response.responseTime).to.be.below(200);
});
```

Let us now add an assertion on the response header using the `Response headers: Content-Type header check` snippet. This can be further modified to check the presence of any header in the response, as shown in the following code snippet:

```
pm.test("Last-Modified is present", function () {
pm.response.to.have.header("Last-Modified");
});
```

Figure 7.14 shows a summary of the test results for the `GET` repository API call with a test status for each of the snippets we have added:

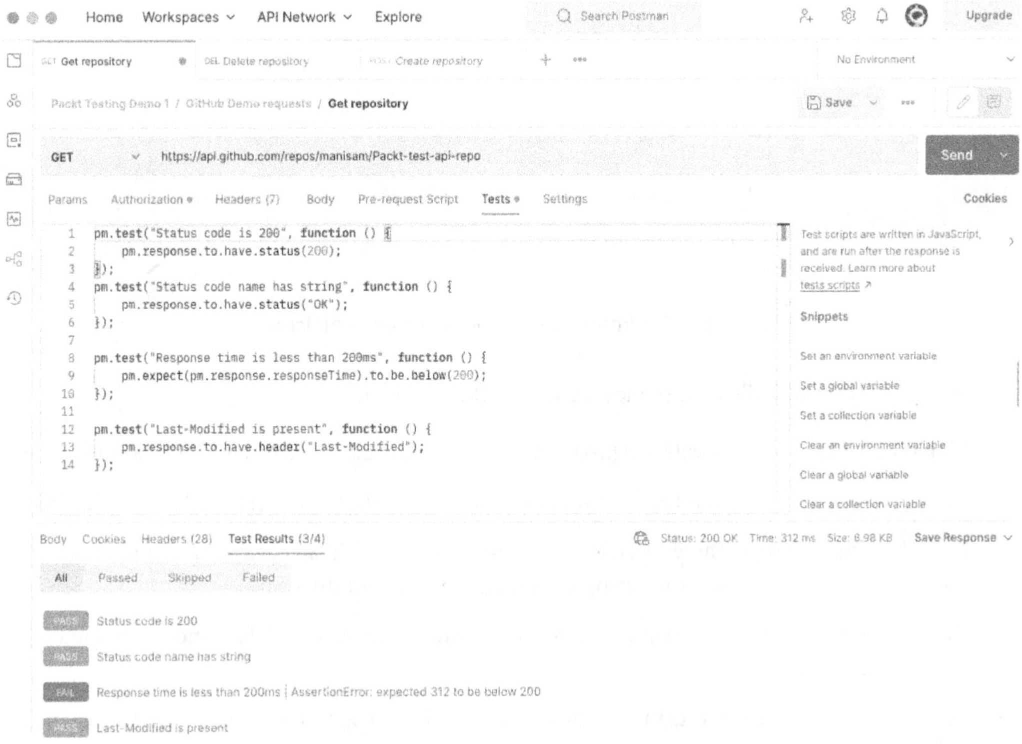

Figure 7.14 – Automated API response validation

These are some of the generic tests we can add for API responses. In the next section, let us learn how to add assertions on specific values such as the repository name on the response.

Understanding Postman variables

Let us understand what an environment means in Postman before jumping into looking at variables. An environment is an assembly of variables that can be used in API requests. For example, multiple environments can be added in Postman, each with its own collection of variables. A new environment can be created and added using the **Environments** drop-down menu and the icon from the top-right section of the workspace window, as shown in *Figure 7.15*. Here, we have created a new **Packt testing** environment:

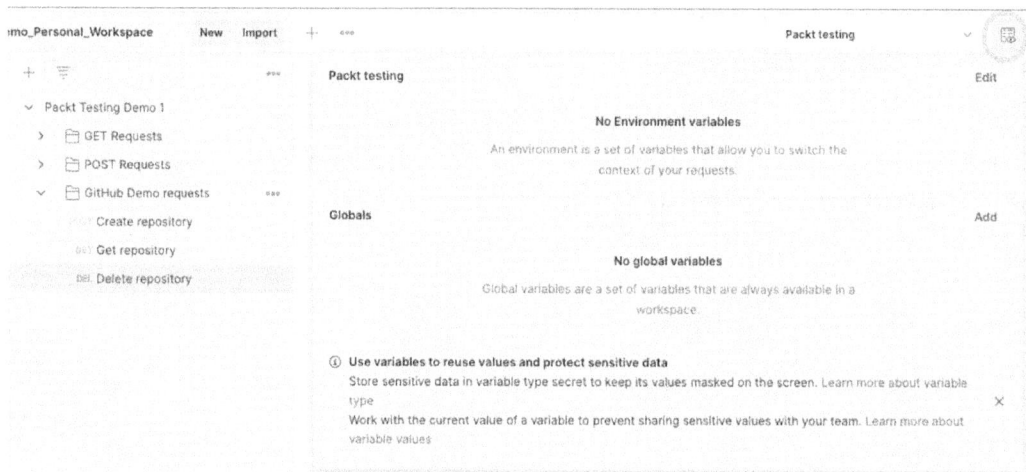

Figure 7.15 – Environments drop-down menu and icon

Postman has five types of built-in variables, which are the following:

- **Global variables**: Variables with the broadest scope and can be accessed anywhere in a workspace

- **Collection variables**: Variables scoped to be available for all the requests within a collection

- **Environment variables**: These variables are accessible only within an environment and primarily used to manipulate access—for example, in staging versus production

- **Data variables**: These are variables created when importing external data and are scoped for a single execution

- **Local variables**: Used on tests in a single request and lose scope as soon as the execution ends

Let us consider the API for creating a new GitHub repository to understand how variables can be used in Postman to remove static context from API requests. For example, we must validate that the repository name matches between request and response. It is also important to remember that GitHub repositories cannot have duplicate names. So, in order for this to work, we should provide a randomly generated name in our POST request body and validate the presence of this name in the corresponding response body. Dynamic variables come in handy to achieve this in Postman. Variables are defined using double curly braces: {{<variable name>}}. The request body for the create repository POST call will look like this:

```
{
 "name": "{{repository_name}}",
 "private": false
}
```

Now, we need this value to be new for every POST request we send, and the best place to do this is in the **Pre-request Script** tab of the request dashboard. This represents a pre-condition to the test case, and it is run before the request is sent to the server. We accomplish random value generation by defining a variable with static and dynamic parts. Here, we use the variables property of the pm object. Then, we set this as an environment variable. *Figure 7.16* shows a new repository_name environment variable created in the current environment, **Packt testing**:

```
let repository_name = "test_packt_api_" + pm.variables.
replaceIn('{{$randomInt}}');
pm.environment.set("repository_name", repository_name);
```

The following figure shows the repository_name environment variable:

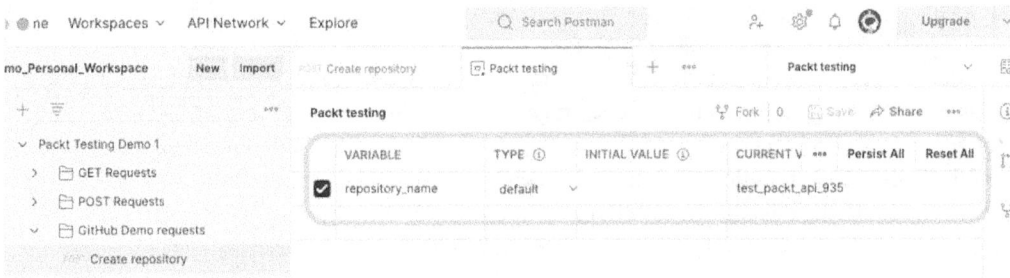

Figure 7.16 – A new environment variable created

> **Note**
> Pre-requisite scripts are run before the API request is executed, while the test scripts are run after the server returns a response.

So far, we have created an environment variable in the **Pre-request Script** tab and updated the request body to use that variable. Let us next add an assertion on this variable in the response body. This can be done in the **Tests** tab via the `Response body: JSON value check` snippet. This snippet helps check a specific value in the API response. Note that the value of an environment variable can be fetched using `pm.environment.get("repository_name")`:

```
pm.test("Check repository name", function () {
var jsonData = pm.response.json();
pm.expect(jsonData.name).to.eql(pm.environment.get("repository_
name"));
});
```

In the next section, let us learn to chain a series of API requests by passing data from one API to the next.

Chaining API requests

Postman allows us to use variables to enable the chaining of a series of API requests. A variable created from the response of an API can be used in the subsequent request. Let us take the example of the create repository call where a new repository is created for every request. The name of this repository can be captured and used in a subsequent `GET` call.

For our understanding here, let us chain the create repository call with `GET` and `DELETE` calls. These calls require the name of the repository and owner. In the create repository call from the previous section, we have the `repository_name` variable. Let us now capture the `owner` variable from the response body using the following code:

```
let json = JSON.parse(responseBody);
pm.environment.set("owner", json.owner.login);
```

We use the `JSON.parse()` method to convert `responseBody` into a `JSON` object and then create an environment variable using the `login` key from the response. `GET` and `DELETE` calls both use the `https://api.github.com/repos/:owner/:repo` route. We create requests for each of these API requests with this route, and on saving, Postman automatically creates a new **Path variables** section in the **Params** tab of the request dashboard. We can now substitute the captured environment variables as values, as shown in *Figure 7.17*. In this way, we are chaining the response of the create repository call to the `GET` and `DELETE` calls:

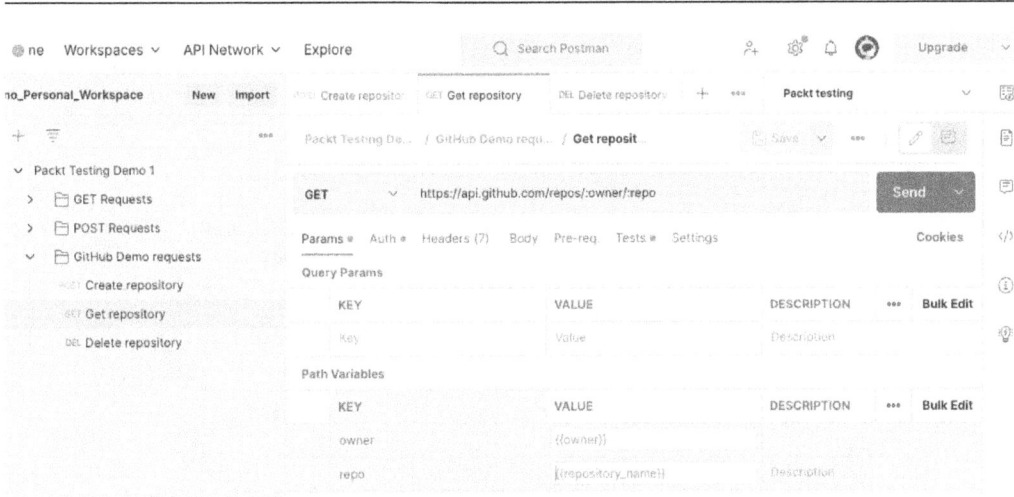

Figure 7.17 – Chaining API requests

On executing these requests one after another, we notice that the values from the request are being passed on to the next one, seamlessly eliminating static data transfer. This behavior makes Postman an effective tool for testing complex API workflows.

We now have a simple collection with multiple API requests, and it is not feasible to run each of them manually for every test cycle. In the next section, let us survey a few ways of executing tests in Postman.

Various ways to execute tests

Let us first look at how to run our tests using the **Collection** runner. This is helpful when all tests in a collection must be run sequentially in an automated fashion. The **Collection** runner window can be launched by using the **Run** button from the **Overview** tab on clicking the collection name. This opens a new tab that displays all the requests in the collection and some additional run parameters. *Figure 7.18* shows this window and the associated options:

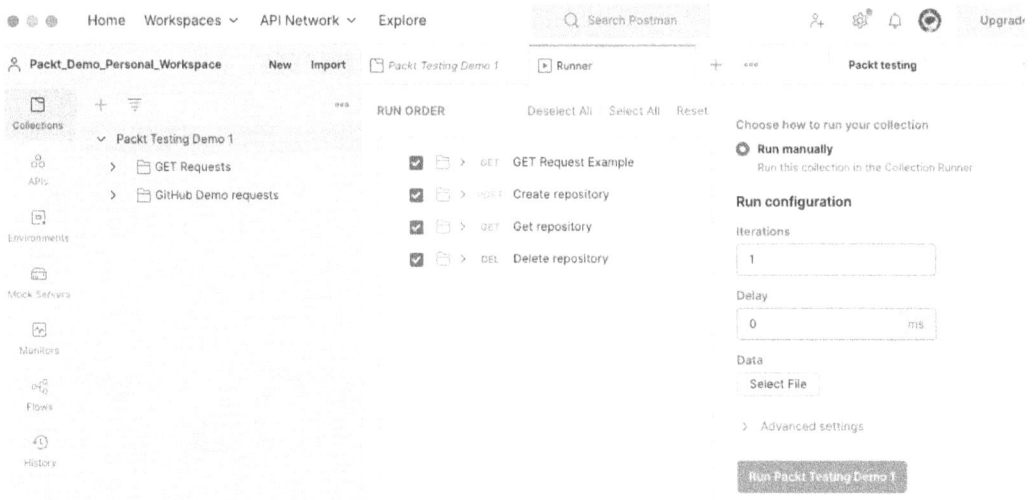

Figure 7.18 – Collection runner window

Using the **Iterations** option, users can specify the number of times the requests would be run. The **Delay** option helps to add a specified wait between subsequent requests. This is very useful in case of long-running requests. There is also an option to upload an external data file and use that data in the form of variables within the request. On clicking the **Run** button, all requests in the collection are run, and results with a clear breakdown are populated in the same window, as shown in *Figure 7.19*:

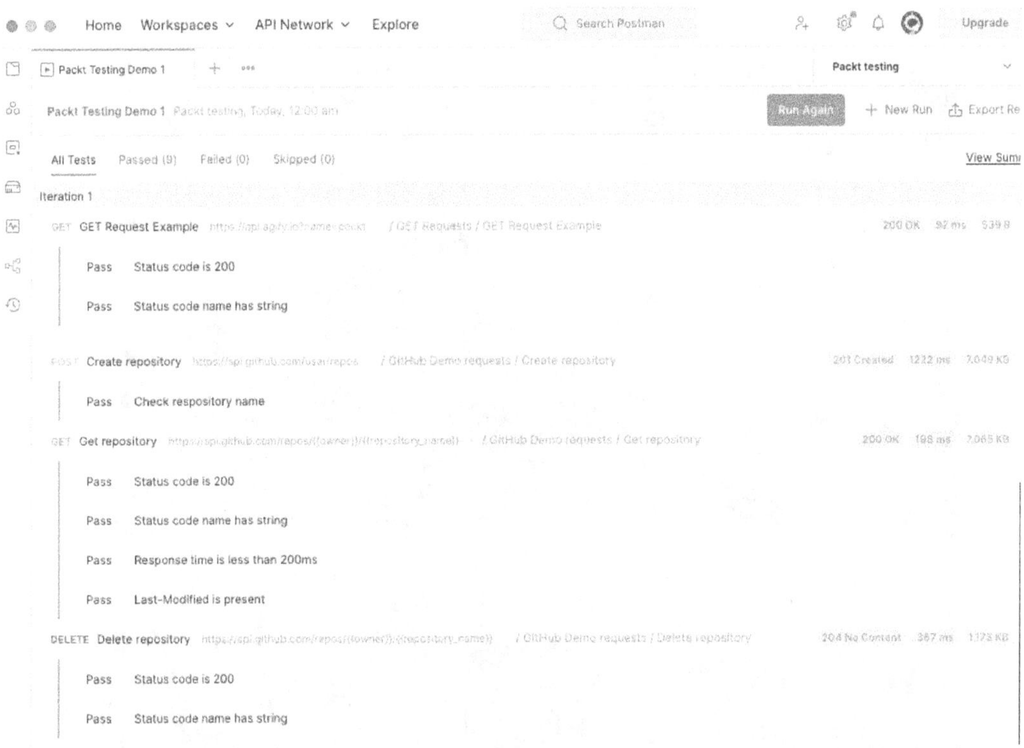

Figure 7.19 – Test results summary

Postman also supports running a collection from the command line through a tool called **Newman**. Newman can be installed using the npm command: `npm install -g newman`. The installation can be verified using the `newman -v` command. To run the collection, we will first export the collections and the associated environment variables. A collection can be exported and saved to the local filesystem using the **Export** option from the **Collections** drop-down menu. Similarly, environment variables can be downloaded by using the **Export** option within **Environments** on the left navigation bar. Note that these files are downloaded in a JSON format. Now that we have the collection and its necessary variables, we can run it using Newman:

```
newman run packt_testing_collection.json -e packt_testing_
environment.json
```

The output from the `run` command line is demonstrated in *Figure 7.20*:

```
→ packt_postman_downloads ls
packt_testing_collection.json   packt_testing_environment.json
→ packt_postman_downloads clear
→ packt_postman_downloads ls
packt_testing_collection.json   packt_testing_environment.json
→ packt_postman_downloads newman run packt_testing_collection.json -e packt_testing_environm
ent.json
newman

Packt Testing Demo 1

□ GET Requests
↳ GET Request Example
  GET https://api.agify.io?name=packt [200 OK, 539B, 620ms]
  ✓ Status code is 200
  ✓ Status code name has string

□ GitHub Demo requests
↳ Create repository
  POST https://api.github.com/user/repos [201 Created, 7.09kB, 1237ms]
  ✓ Check resspository name

↳ Get repository
  GET https://api.github.com/repos/manisam/test_packt_api_131 [200 OK, 7.11kB, 208ms]
  ✓ Status code is 200
  ✓ Status code name has string
  1. Response time is less than 200ms
  ✓ Last-Modified is present

↳ Delete repository
  DELETE https://api.github.com/repos/manisam/test_packt_api_131 [204 No Content, 1.17kB, 377
ms]
  ✓ Status code is 200
  ✓ Status code name has string
```

	executed	failed
iterations	1	0
requests	4	0
test-scripts	4	0
prerequest-scripts	1	0
assertions	9	1

```
total run duration: 2.5s

total data received: 11.39kB (approx)

average response time: 610ms [min: 208ms, max: 1237ms, s.d.: 390ms]

#   failure         detail

1.  AssertionError   Response time is less than 200ms
                     expected 208 to be below 200
                     at assertion:2 in test-script
                     inside "GitHub Demo requests / Get repository"
→ packt_postman_downloads ▌
```

Figure 7.20 – Collection CLI run

Postman provides integration with Docker for executing tests within a Docker container. Docker is a platform that assists in building, deploying, and testing your application code on units called containers, irrespective of the underlying operating system. It provides great portability for developing and testing applications. Running a collection on Postman's Docker container involves just a couple of commands. Once you have Docker installed on your machine, run the `docker pull postman/newman` command. This command pulls the latest image of the Postman `docker/newman` runner from Docker Hub and sets up the container. Next, we need the URL of the collection to be able to run it externally. This can be obtained using the **Share** option from the **Collections** drop-down menu. Now, run the following Docker command:

```
run -t postman/newman run "<<Collection URL>>"
```

This brings us to the end of a basic exploration of API testing with Postman. The Postman tool has so much more to offer, and its capabilities can be referenced at `https://learning.postman.com/docs`.

In the next section, let us review the considerations that go into API automation testing.

Key considerations for API automation

Automating API tests involves a thorough understanding of the underlying business logic as well as the architectural nuances. Data and performance aspects of the API always pose a tough challenge when trying to run multiple iterations of API tests.

Effective API test automation

Let us now look at some chief considerations when automating API tests:

- Static data references must be eliminated wherever possible. These render the tests inextensible and isolated.

- Do more than happy path testing since API tests are quick to run and can provide feedback early in the product life cycle. Applicable product use cases should be tested to a greater extent via APIs. Consider mocking and stubbing to fill in gaps to complete the application flow.

- Since there is no frontend to check the test results, it is imperative to keep API tests clear with sufficient documentation. This helps especially in debugging when there are new failures in the system.

- Manage the data needed for a test by employing setup and teardown tests wherever possible. This helps clean up the environment before and after the execution.

- Pay special attention to authentication mechanisms and try to implement these the same way a real user would access the API.

- Add extra checks for data consistency when chaining responses of multiple services. Passing the same data in different ways may not look bad at the API level but might end up breaking the UI layer of the application.

Throughout this chapter, we examined the requests and responses of REST APIs. As we saw earlier, REST is an architectural framework for creating and accessing APIs. In the next section, let us review and compare the testing aspects of **GraphQL** with REST.

Testing GraphQL versus REST APIs

GraphQL is a query language specification while REST is an architectural framework. Even though one is not considered a replacement for the other, the areas of focus shift when it comes to testing them. Let us now examine them one by one:

- Tests on GraphQL should include schema validation while REST demands endpoint validation.

- After the initial learning curve, changes can be done faster in GraphQL APIs due to its client-driven architecture. Test automation cycles must be designed to keep up with these rapid changes.

- Validations on API output should be customized to the specific GraphQL query being tested. This is different in the case of REST with its fixed response on endpoints.

- GraphQL was designed to ease API integration bottlenecks, while REST provides a solid foundation for individual API designs. A higher effort may be involved in testing individual GraphQL APIs while REST simplifies isolated API testing.

GraphQL is a very detailed specification on its own, and readers are encouraged to go through the documentation for further exploration.

This brings us to the end of this chapter. Let us now quickly summarize what we learned in this chapter.

Summary

We acquired a basic understanding of REST APIs and went through a basic setup of Postman with workspaces. We looked in detail at sending GET/POST requests and how to validate their responses. We then learned what collections are and how they help organize API requests within our workspace for conciseness and collaboration. In the next section, we learned how to validate responses in an automated fashion using snippets. We gained an understanding of how variables are used in Postman to chain API requests. Subsequently, we surveyed various ways to execute collections. Finally, we took up some key considerations that go into automating API tests and familiarized ourselves with how GraphQL APIS are different from their REST counterparts.

In the next chapter, we dive deep into performance testing with the JMeter tool.

Questions

1. What are common areas to focus on in API testing?

2. What are collections in Postman?

3. What are snippets and how do they help?

4. How can complex workflows be tested in Postman?

5. What is Newman used for?

6. What is Docker and how does it help?

7. What is GraphQL and how is it used?

8
Test Automation for Performance

Performance testing ensures the speed, stability, and scalability of a software application and helps to verify the behavior of an application in various possible user scenarios. Even with considerable technological advancements, many software applications fail to produce the desired performance for users during peak traffic. Therefore, it is imperative that performance testing is done early enough in the application life cycle to provide critical feedback to the stakeholders and to maintain usability at elevated levels.

There are various performance testing tools on the market to choose from, and each one has its own capabilities. Some examples are **JMeter**, **LoadRunner**, and **Kobiton**. In this chapter, we will be working hands-on to gain an understanding of setting up and running performance tests using JMeter. These are the main topics we will be looking at in this chapter:

- Getting started with JMeter
- Automating a performance test
- Java essentials for JMeter
- Considerations for performance testing

Technical requirements

To get functional with JMeter, we need Java installed on our machine. Currently, JMeter works with JDK 8 and JRE 8 or higher.

Getting started with JMeter

Let's begin with a basic understanding of JMeter, what it does, and how to install it.

What is JMeter and how does it work?

JMeter is a performance-testing application built using Java. It is a completely free and open source tool created by Apache, and it can be used to performance test a wide variety of applications, including APIs and databases. JMeter's capabilities can be extended by a range of plugins that it supports. JMeter simulates multiple user load on the test application and sends requests to the server. When the server responds to each of the incoming requests, JMeter collects the response and the associated stats and displays them in a report utilizing tables and graphs. The **test plan** is one of the primary components of JMeter, and it comprises a user-configured series of steps to be run as part of a load test. JMeter provides options to execute a test plan via a GUI or through the command line.

Let's start by downloading and installing JMeter.

Installing JMeter

Let's look at the steps involved in the installation of JMeter:

1. The first step in the installation of JMeter is to check the Java version on your machine. This can be done using the `java -version` command. As shown in *Figure 8.1*, this command prints out the JDK version:

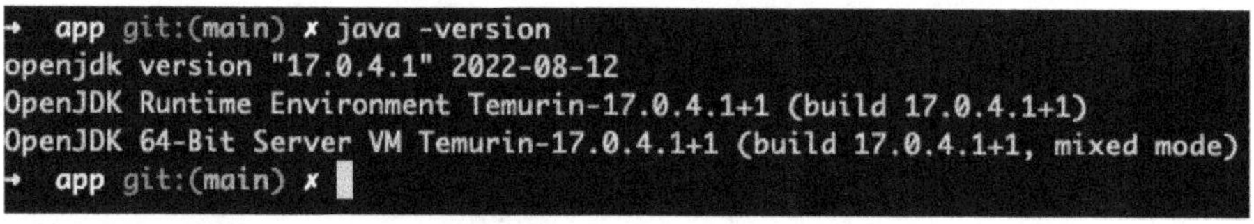

```
→  app git:(main) x java -version
openjdk version "17.0.4.1" 2022-08-12
OpenJDK Runtime Environment Temurin-17.0.4.1+1 (build 17.0.4.1+1)
OpenJDK 64-Bit Server VM Temurin-17.0.4.1+1 (build 17.0.4.1+1, mixed mode)
→  app git:(main) x ▌
```

Figure 8.1 – Checking the Java version

2. The next step is to download the JMeter binary. The file for download can be found on the JMeter website, `https://jmeter.apache.org/download_jmeter.cgi`. In this case, I am downloading the binaries zip file for version 5.5, as shown in *Figure 8.2*.

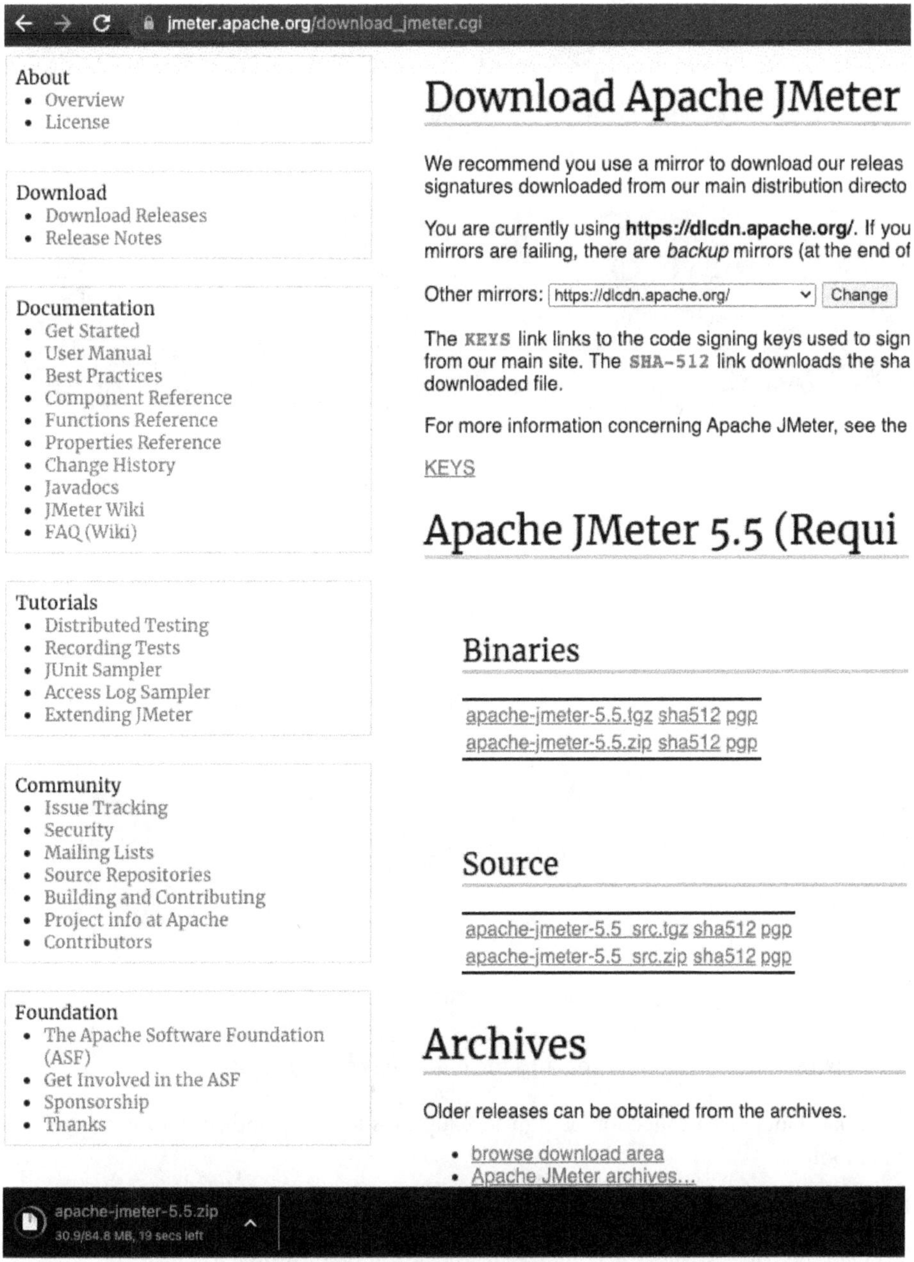

Figure 8.2 – JMeter downloads page

3. Once the download is complete, move the zipped file to the desired local folder and unzip it. This should create a new folder in the same location within which all the contents are extracted.

JMeter can now be started with the `sh jmeter.sh` command from the `bin` folder of the application. This brings up the application in a separate window, as shown in *Figure 8.3*.

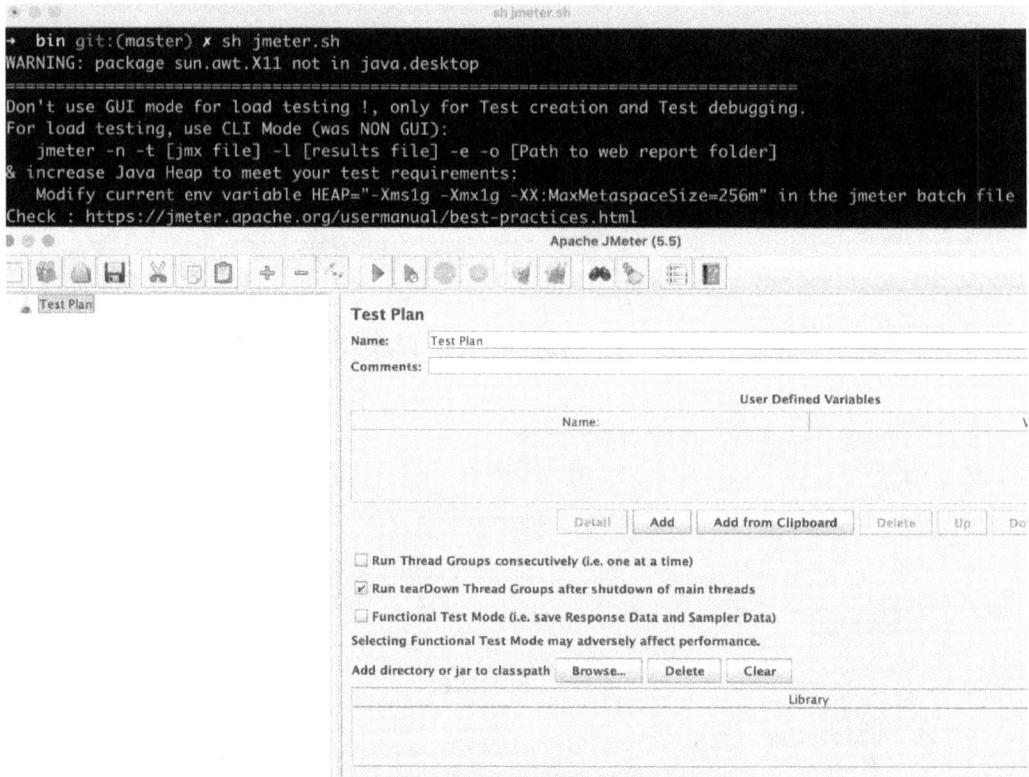

Figure 8.3 – Starting JMeter

JMeter comes with a simple GUI that contains the following components:

- Menu bar: Contains a collection of high-level options to set up and configure various aspects of the tool

- Tool bar: Contains frequently used tools

- Test plan tree view: Groups all components that are added within a test plan

- Editor section: Provides options to edit the selected component from the test plan view

In the next section, let's look at how to create our first performance test in JMeter.

Automating a performance test

JMeter provides an intuitive GUI that we can use to create and configure performance tests. The test plan is at the core of a performance test, and we start by creating one. We can do this by either using the **New** option from the menu bar or the tool bar. We looked at the new **Test Plan** window in *Figure 8.3* when launching the JMeter application.

Building and running our first performance test

One of the primary focuses of a performance testing tool is its ability to simulate multiple users. This is accomplished in JMeter by configuring a thread group. As shown in *Figure 8.4*, this is done via the **Thread Group** option under **Test Plan**.

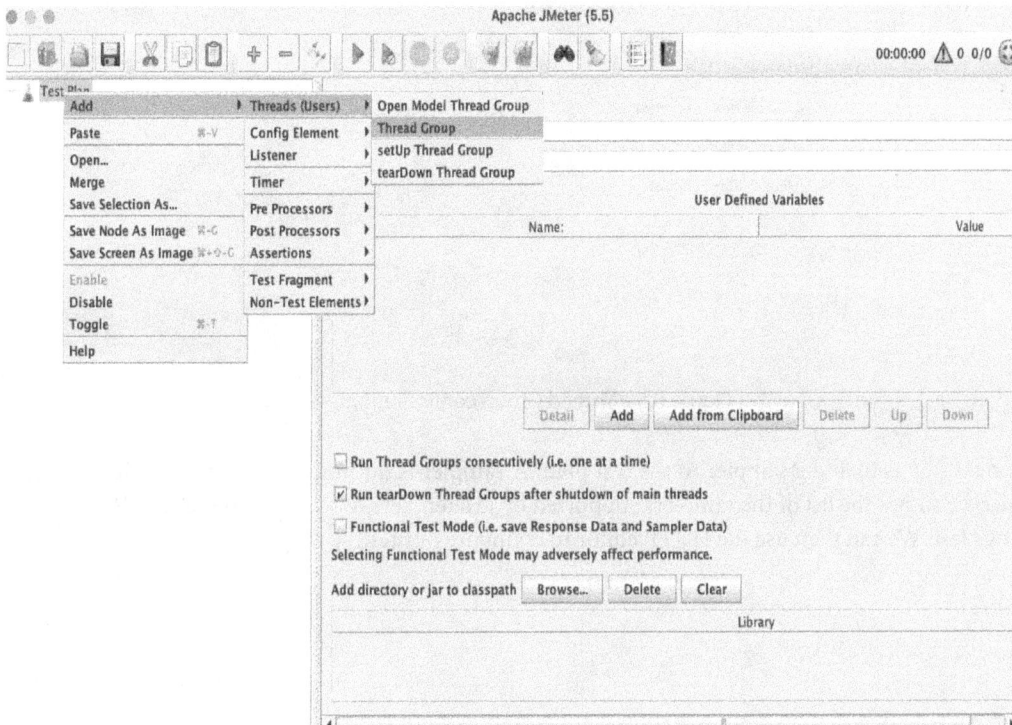

Figure 8.4 – New thread group

We use a combination of three parameters to achieve the required pacing for our performance test:

- **Number of Threads**: The number of parallel users to be simulated in this test
- **Ramp-up period**: The time taken to simulate the specified number of users
- **Loop Count**: The number of iterations to be executed as part of the current test

There are additional settings to configure and fine-tune the load on the test, as shown in *Figure 8.5*.

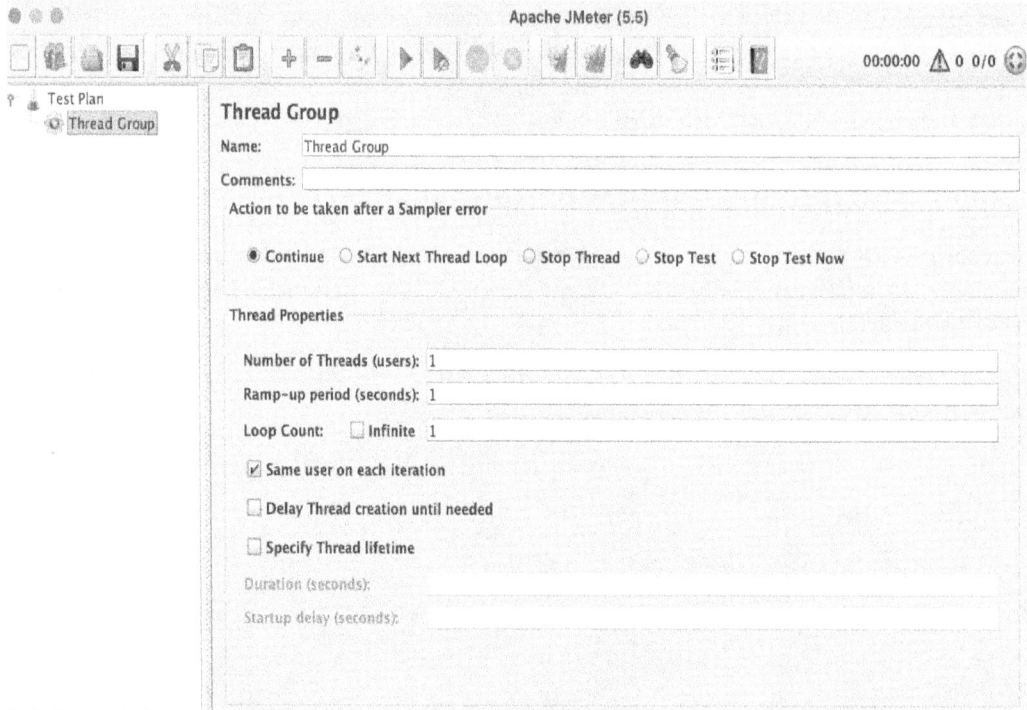

Figure 8.5 – Thread Group configuration

The next step is to add a sampler to the test plan. A sampler is nothing but a test added in JMeter. *Figure 8.6* shows the list of the samplers supported by JMeter. Let's now add a simple HTTP sampler for our test. We can then use the HTTP editor to configure our test.

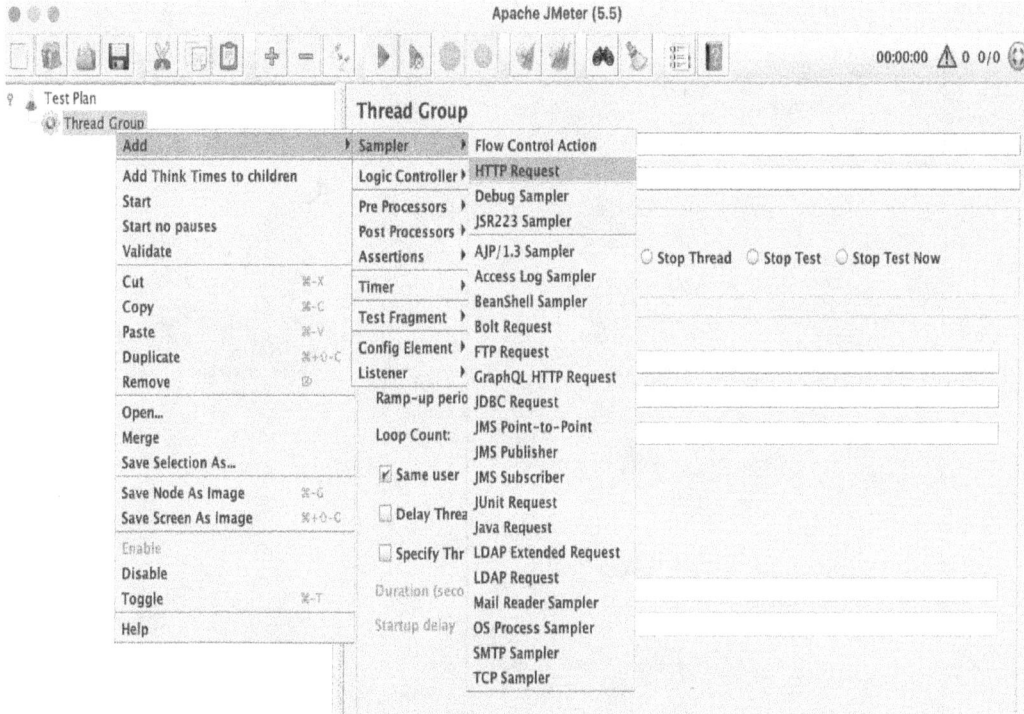

Figure 8.6 – Adding an HTTP sampler

In our example, we will be load-testing the Packt Publishing website at `https://packtpub.com`. The URL is split between the fields, protocol, and server name in the editor. Then we specify the path, `/terms-conditions`, in the **Path** field. We will be testing the GET request, but there are in-built options to support other types of requests, along with request body and file uploads, as shown in *Figure 8.7*.

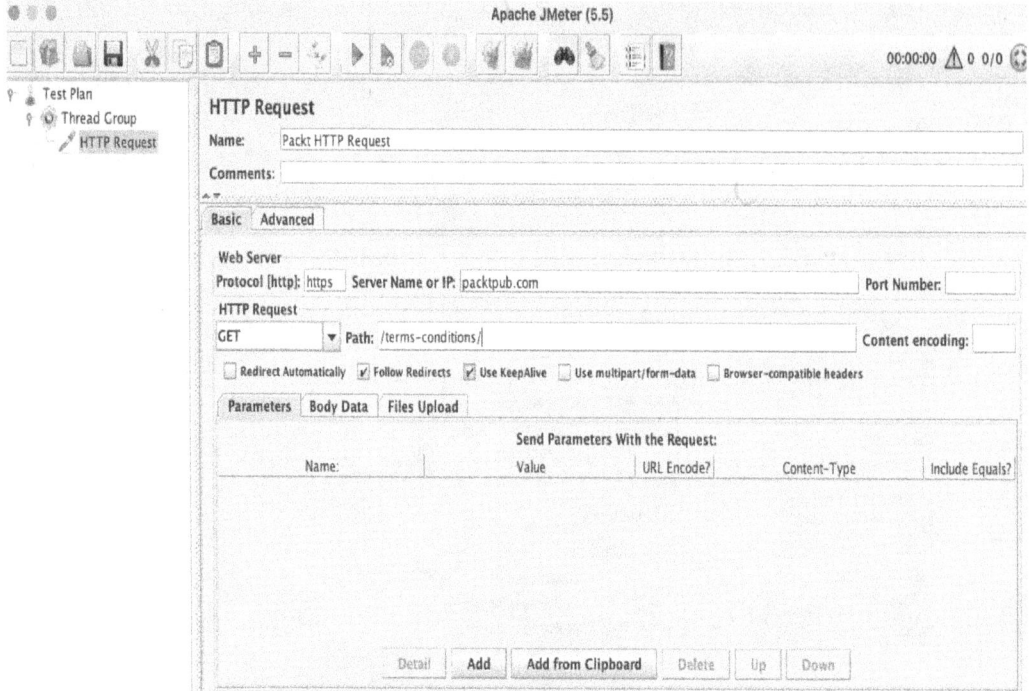

Figure 8.7 – Configuring an HTTP sampler

The test plan can now be saved to a local directory using the **Save** option from the menu bar. The next step is to add a **listener**, which helps us view the test results. A listener is a component within a test plan that stores and allows us to views results. Let's add the **View Results Tree** and **View Results in Table** listeners to our test plan. JMeter provides a variety of listener options in the **Add | Listener** menu. *Figure 8.8* shows our test plan with the listeners added.

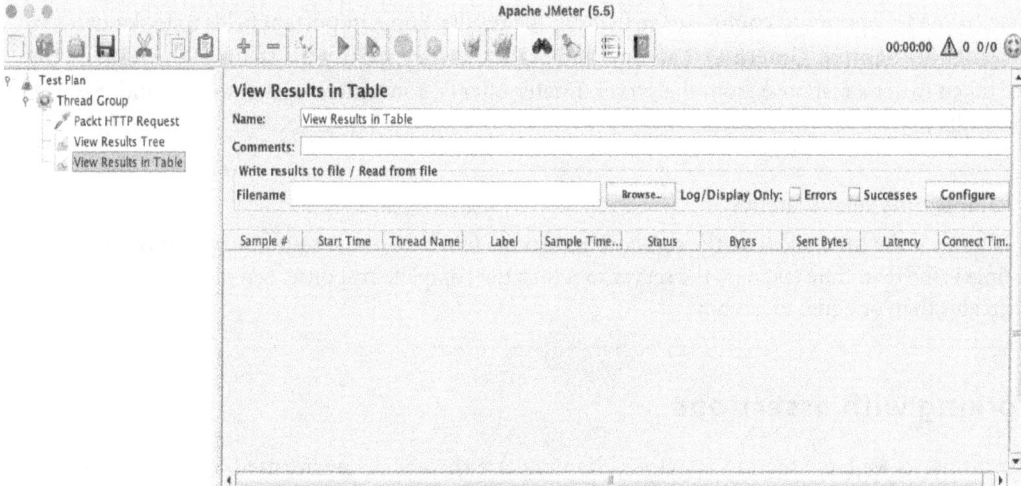

Figure 8.8 – Listeners in a test plan

After saving the test plan, we are ready to execute our first test. This is done using the **Start** button on the menu bar. We can see the results being populated in the listeners as soon as the test begins. *Figure 8.9* shows a breakdown of the test run stats by the thread group within the **View Results in Table** listener.

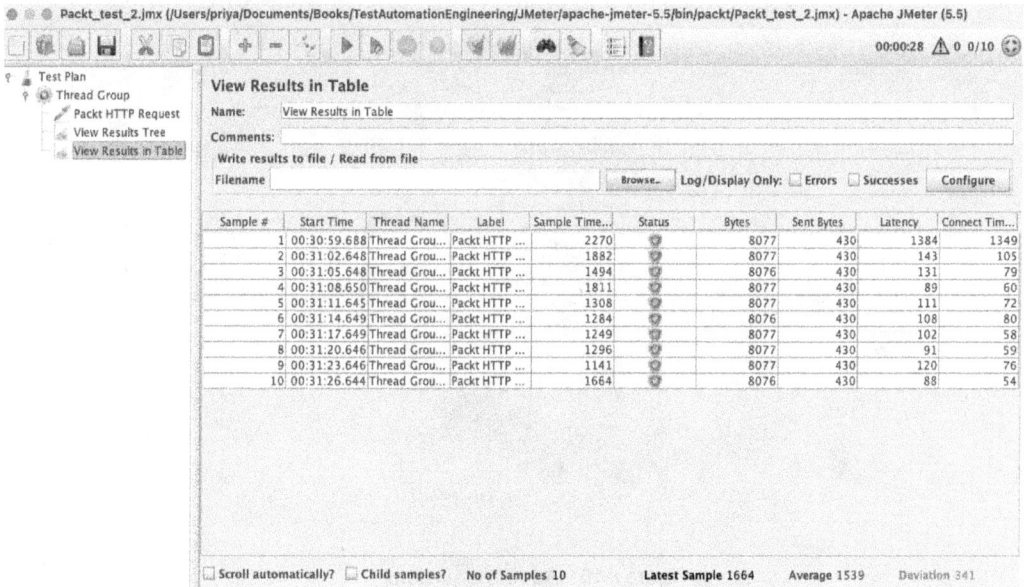

Figure 8.9 – Test run results

JMeter provides options to configure the fields in the results. Some important fields to look out for in the results are **Sample Time(ms)**, **Latency**, and **Status**, as they specify the status of the test and the time taken to get a response from the server. JMeter offers a convenient option to save and view the test results in CSV or XML format.

> **Sample time versus latency**
>
> Latency is the time taken by the server to return the first byte of the response, whereas sample time is the total time taken by the server to return the complete response. Sample time is always greater than or equal to latency.

Working with assertions

Assertions, as we have seen in previous chapters, are the checks performed on the request and response. JMeter provides options to perform checks on an array of options, such as response size, response time, the structure of the response, and so on. An important thing to note about assertions is that they can be added at all levels. For example, an assertion added at the test plan level will apply to every sampler within it. For our example, let's add response and duration assertions for the HTTP sampler, as shown in *Figure 8.10*.

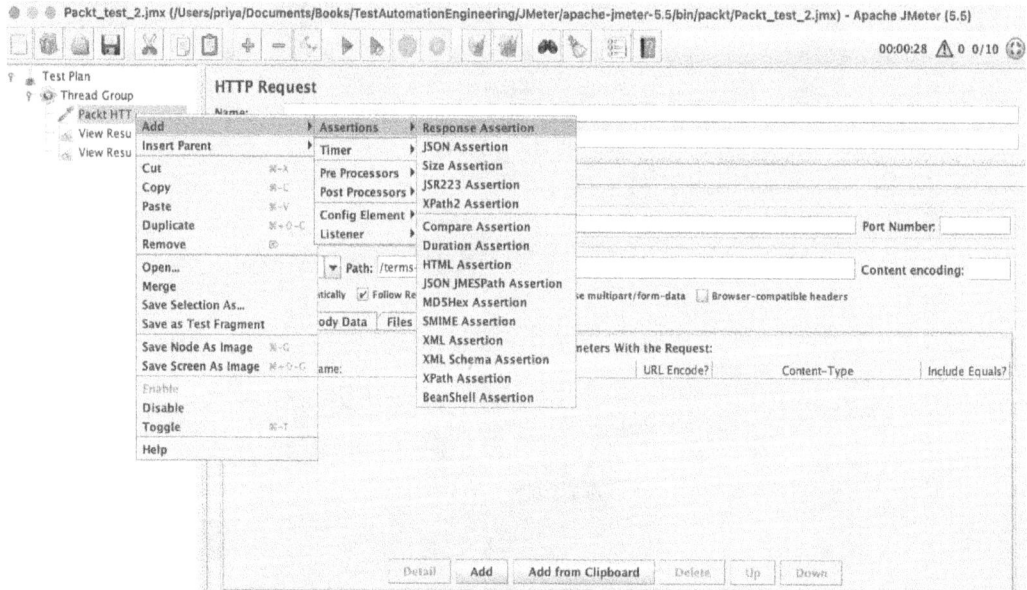

Figure 8.10 – Adding assertions

Let's update the **Response Assertion** to look for the response code 200 and the **Duration Assertion** to flag responses over 1,000 ms. These conditions are checked after every iteration of the HTTP sampler, and the results are flagged accordingly. *Figure 8.11* demonstrates the execution of the **Duration Assertion** where some of the responses took over a second to complete.

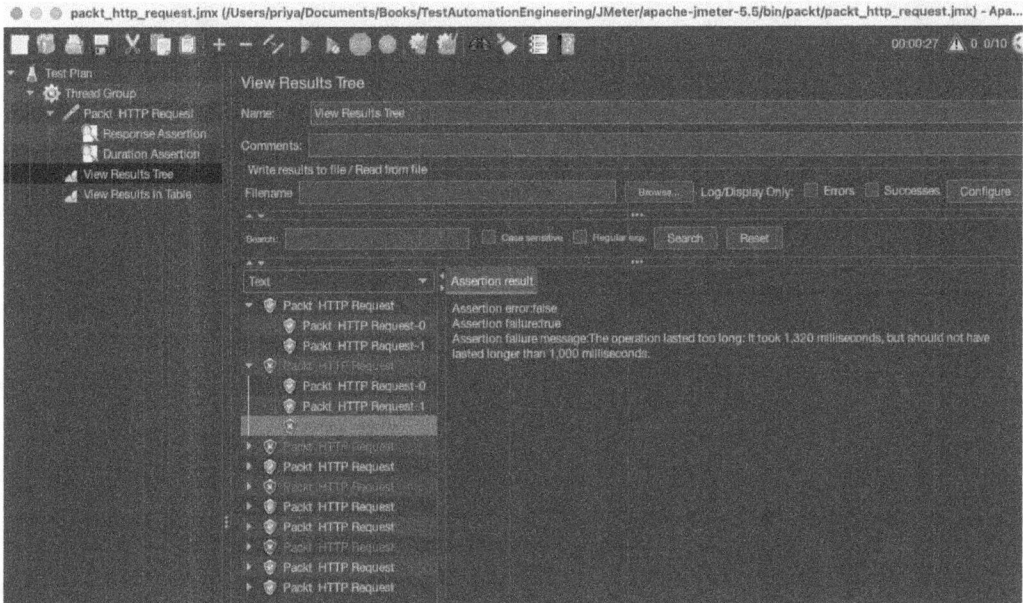

Figure 8.11 – Assertion results

The **Assertion Results** listener is an effective component that collates the responses from all the assertions so you can view them in one place. This listener can be added at the test plan level, as illustrated in *Figure 8.12*. It combines the results from the **Response Assertion** and the **Duration Assertion**.

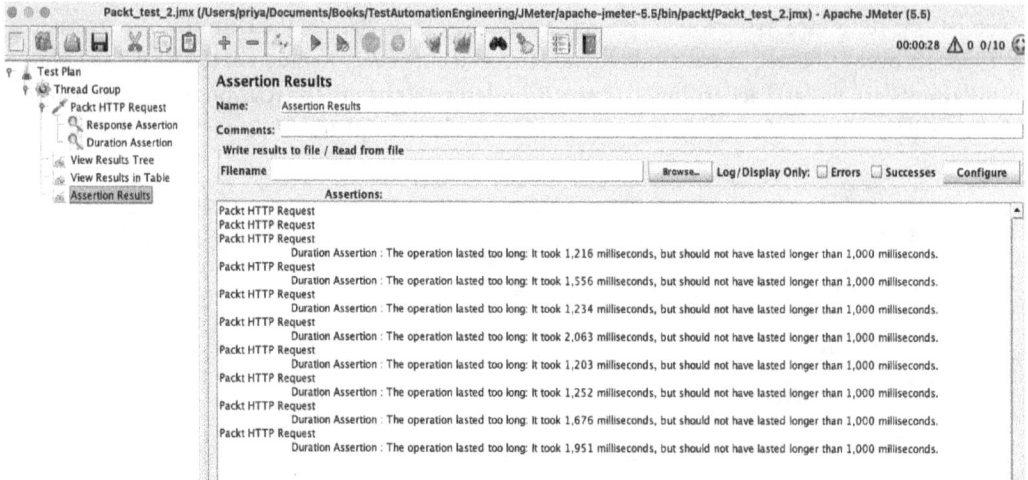

Figure 8.12 – Assertion results listener

Let's now look at how to use the command line to handle JMeter's tests.

Working with tests via the command line

Performance tests are often long-running and tend to be heavy on system resource consumption. GUI mode consumes a lot of memory, especially when running pre-recorded scripts, and execution via the command line alleviates this pain by reducing the memory footprint of the tool. Another significant benefit is the ability of the command line to integrate easily with external processes, such as continuous integration systems. In this section, we will learn how to configure and run a JMeter test from the command line.

Let's reuse the test plan from the previous section for execution via the command line. Navigate to JMeter's `bin` folder in your command line and run the following command:

```
sh jmeter -n -t "<location of the .jmx test plan>"-l "<location
to log the results>"
```

Here, -n stands for non-GUI mode, -t specifies the location of the test plan, and −l is the location of the result logs. The command line supports various other parameters, but these are the minimum required parameters to trigger the execution. *Figure 8.13* shows the execution of a command line run.

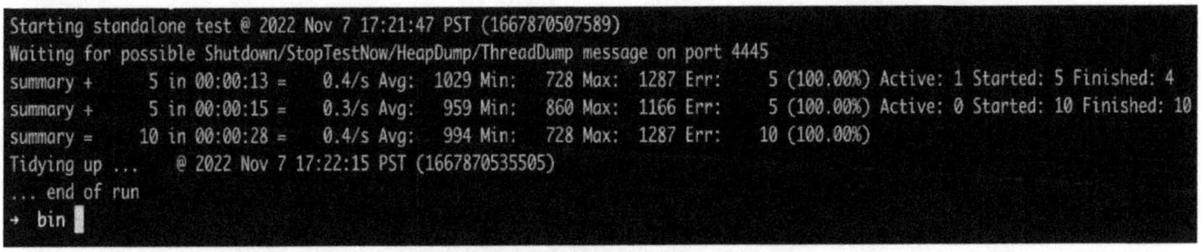

```
Starting standalone test @ 2022 Nov 7 17:21:47 PST (1667870507589)
Waiting for possible Shutdown/StopTestNow/HeapDump/ThreadDump message on port 4445
summary +     5 in 00:00:13 =   0.4/s Avg:  1029 Min:  728 Max:  1287 Err:    5 (100.00%) Active: 1 Started: 5 Finished: 4
summary +     5 in 00:00:15 =   0.3/s Avg:   959 Min:  860 Max:  1166 Err:    5 (100.00%) Active: 0 Started: 10 Finished: 10
summary =    10 in 00:00:28 =   0.4/s Avg:   994 Min:  728 Max:  1287 Err:   10 (100.00%)
Tidying up ...   @ 2022 Nov 7 17:22:15 PST (1667870535505)
... end of run
→ bin ▐
```

Figure 8.13 – JMeter command line run

Additionally, the sh jmeter -h command can be used to review all the available command-line options.

Performance test results can get voluminous, and it is always necessary to produce a clear and concise report. It will be hard to understand the test results with just the results shown on the command line and it necessitates a better report. This is achieved by using the −e option, which generates a dashboard report, and the −o option to specify the location of the results folder. *Figure 8.14* shows a part of the HTML report generated when using these parameters. By default, this is produced as an index.html file within the results folder specified as part of the command-line option. The full command to achieve this is as follows:

```
sh jmeter -n -t "./packt/packt_http_request.jmx" -l "./packt/
report.csv" -e -o "./packt/dashboard_report"
```

Apache JMeter Dashboard

🏠 Dashboard	
📊 Charts	<
📊 Customs Graphs	<

Test and Report information

Source file	"report.csv"
Start Time	"11/7/22, 5:55 PM"
End Time	"11/7/22, 5:55 PM"
Filter for display	""

APDEX (Application Performance Index)

Apdex	T (Toleration threshold)	F (Frustration threshold)	Label
0.583	500 ms	1 sec 500 ms	Total
0.000	500 ms	1 sec 500 ms	Packt HTTP Request
0.850	500 ms	1 sec 500 ms	Packt HTTP Request-0
0.900	500 ms	1 sec 500 ms	Packt HTTP Request-1

Requests Summary

FAIL
33.33%

PASS
66.67%

FAIL
PASS

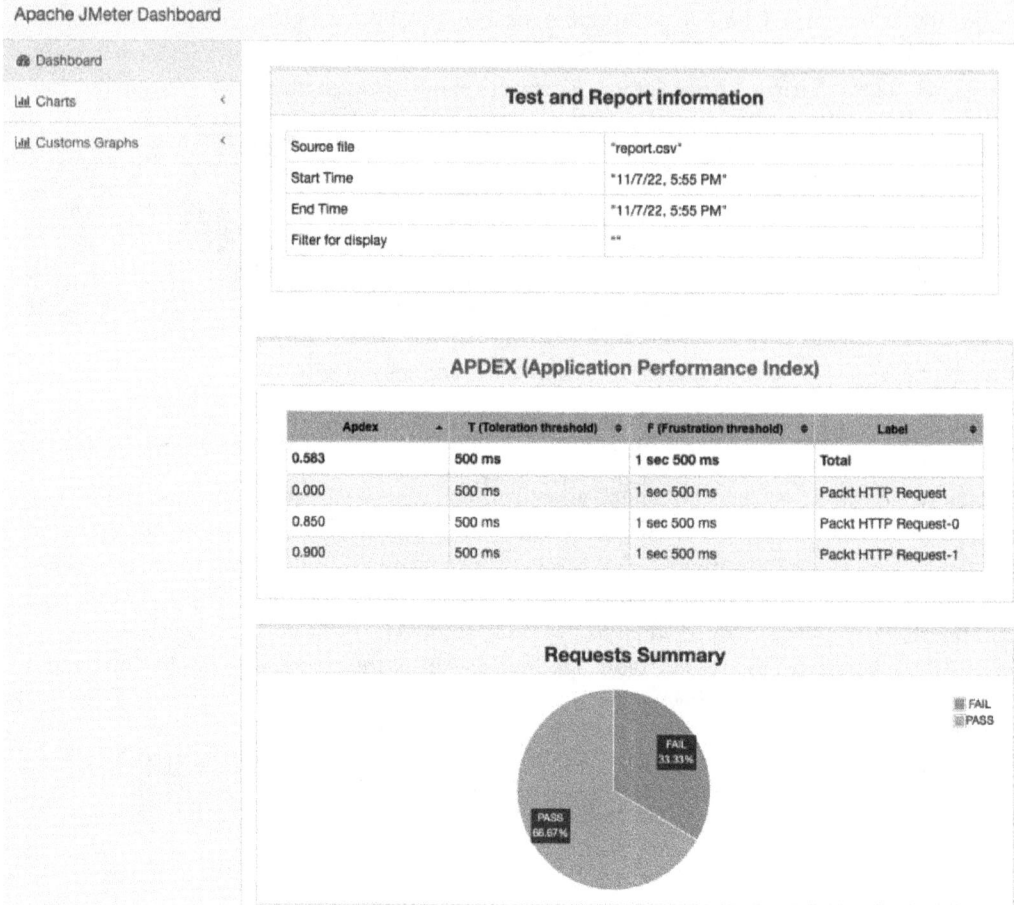

Figure 8.14 – JMeter Dashboard report

Another powerful feature that the JMeter command line provides is the use of built-in functions to send dynamic parameters when running a test plan. For example, in our test plan, we have hardcoded the path as /terms-conditions. In real time, we would be testing different paths from the command line and would not have to update the test plan for every run. The test plan can be updated with a function in this field to be able to receive this parameter via the command line using the format ${__P(VariableName)}. The path can now be sent through the command line by prefixing J to the variable name:

```
sh jmeter -n -t "./packt/packt_http_request.jmx" -Jpath=/terms-
conditions
```

In the next section, let's look at how to use the **HTTP(S) Test Script Recorder** component in JMeter.

Using the HTTP(S) Test Script Recorder

The **HTTP(S) Test Script Recorder** is a component that records requests from the browser. Previously, we manually added the HTTP request, but this component adds them automatically by recording the transactions. This option can be added directly under the test plan, as shown in *Figure 8.15*.

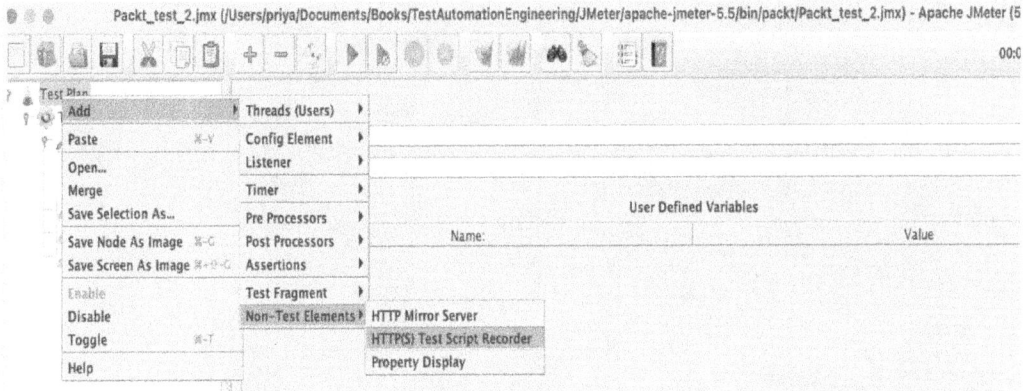

Figure 8.15 – Adding the HTTP Test Script Recorder component

We will also need a recording controller to be added to the test plan to categorize the recording by the traffic or per page. For simplicity, let's use a single controller here, but in real-world scenarios where the user flows involve multiple pages, we might need a separate controller per page. The **Target Controller** property should be set to point to the right controller within the **HTTP Test Script recorder**. Another notable feature within the **HTTP Test Recorder** component is **Request Filtering**. A lot of resources are exchanged when recording HTTP requests, and not all of them will be applicable for load testing. URL patterns that need to be included or excluded can be specified using the **Request Filtering** option.

The next step is to configure the proxy on our browser so that only the desired traffic flows through the port. This is done by specifying the default JMeter port 8888 within the browser's proxy configuration. *Figure 8.16* shows this configuration on the Chrome browser.

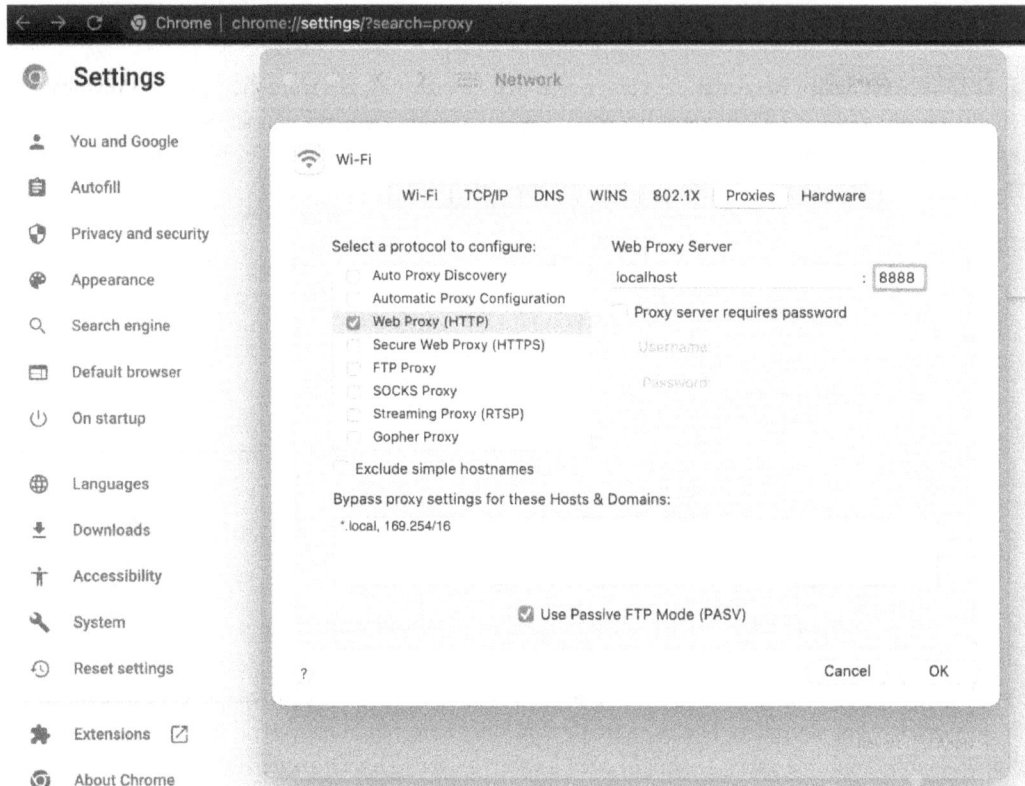

Figure 8.16 – Chrome proxy configuration

There is one more step before we can start recording, and that is to add the JMeter certificate to the browser. This file (`ApacheJMeterTemporaryRootCA.crt`) can be found in JMeter's bin folder, and it needs to be added to the browser certificates via settings. Once this is done, we can use the **Start** button on the recorder component to commence the recording. When the recording is complete, the HTTP requests are stored under the corresponding controller. These requests can then be played back with the simulated load.

We have gained foundational knowledge on how JMeter operates, and we recommend you to further explore the tool using the user manual at `https://jmeter.apache.org/usermanual/index.html`. Let's move on to the next section to gain a basic understanding of the Java programming language and how to use it to write custom code within JMeter.

Java essentials for JMeter

There may be instances where the features that come out of the box with JMeter are not sufficient and custom scripts are needed to perform specific tasks. JSR233 and Beanshell assertions/samplers can be utilized in cases such as these to get the job done. Both these components support Java code, and hence it is important to acquire basic Java knowledge. In this section, let's go through a quick introduction to the Java programming language.

A quick introduction to Java

Java is a platform-independent compiled programming language. Java code gets compiled into bytecode, which can then be executed on any OS. The **Java Virtual Machine (JVM)** is the OS-specific architectural component that sits between the compiled bytecode and the OS to make it work on any platform. Let's now create our first Java program, compile it, and run it. Any Java program comes with a boilerplate code, as follows:

```
package ch8;
public class first_java_program {
    public static void main(String[] args) {
    }
}
```

Let's familiarize ourselves with these keywords, to begin with. Whenever a new class is created in Java, the very first line is usually the package name, followed by the class definition. The `public` keyword is an access modifier that denotes the access level of this class. This is followed by the `class` keyword and the name of the class. Within the class, there is always a `main` method with a `public` access modifier.

This is followed by another keyword, `static`, which signifies that this method can be invoked directly without the need to create an instance of the class. The `main` method is always called by the JVM at the beginning of program execution, and that is why we do not need an instance of the class to call this method. Next is `void`, which represents the return type of this method. In this case, we do not return anything and hence leave it as `void`. We could return a string or an integer depending on what is being done within the method. The values within the parentheses after the method name mark the arguments accepted by the method.

Let's add a simple `print` statement, `System.out.println("My first java program")` within the `main` method and run it via the IDE. This should print the text specified within the `println` method. This completes our first program in Java. Java is a strong object-oriented language, so let's learn how to create classes and objects in Java.

Object-oriented programming primarily helps us model real-world information in our programs. Let's take an example of a bank account and see how it can be modeled in Java using object-oriented techniques. To start with, let's create an `Account` class, as follows:

```
package ch8;
```

```
public class Account {
String account_holder_name;
int age;
float account_balance;
boolean direct_deposit_enabled;
Boolean maintains_minimum_balance;
public void test_minimum_balance(){
if (account_balance > 5000) {
maintains_minimum_balance = true;
}
}
}
```

This class would act as a template for all the accounts that are created. Each account created from this template would be an object, or an instance, of this class. We have used different variable types to model real-world information. We have also used the test_minimum_balance method to derive and set the value of a variable called maintains_minimum_balance within the class.

Let's now go ahead and create another class that holds these objects:

```
package ch8;

public class AccountObjects {
public static void main(String[] args) {
Account johns_account = new Account();
Account davids_account = new Account();

johns_account.account_holder_name = "John Doe";
johns_account.age = 32;
johns_account.account_balance = 10000;
johns_account.direct_deposit_enabled = true;
johns_account.test_minimum_balance();

tims_account.account_holder_name = "Tim Sim";
tims_account.age = 35;
tims_account.account_balance = 1000;
tims_account.direct_deposit_enabled = true;
tims_account.test_minimum_balance();
```

```
    }
}
```

We have created two objects in our second class, which represent two different people's bank accounts. This example demonstrates how we can use classes to model information.

> **JDK versus JRE versus JVM**
>
> **JDK**: The **Java Development Kit** is an environment for developing, compiling, and running Java applications.
>
> **JRE**: The **Java Runtime Environment** is an environment for running Java applications. Users of Java applications just need the JRE.
>
> **JVM**: The **Java Virtual Machine** is an interpreter for executing Java programs.

This section was meant to quickly inform you what the Java programming language is and how to write a basic program. You are encouraged to refer to the official Java documentation at `https://docs.oracle.com/javase/tutorial/getStarted/index.html` to further your knowledge. Let's now get back to writing custom scripts in JMeter.

Using the JSR233 assertion

JMeter comes with a JSR233 assertion/sampler that can interpret and execute Java code. JSR233 is a scripting API for languages that can work on the JVM. Apache Groovy, Python, and Ruby are some of the supported languages, and we will be using Groovy for our example as it provides better performance.

Another advantage of using Groovy is that it is an extension of the JDK and accepts Java code. In fact, it supports all the features of Java and provides additional dynamic features, whereas Java is a strongly typed language. Groovy's official documentation can be found at `https://groovy-lang.org/documentation.html`. Since the Groovy engine is part of JMeter, no additional installation is required to get it working. Let's now look at how to employ a Groovy script within a JMeter test plan using the JSR233 sampler/assertion.

To start, let's add the JSR233 assertion to the HTTP request in our existing test plan. By default, Groovy is selected as the language for this assertion, but there are other options, as shown in *Figure 8.17*.

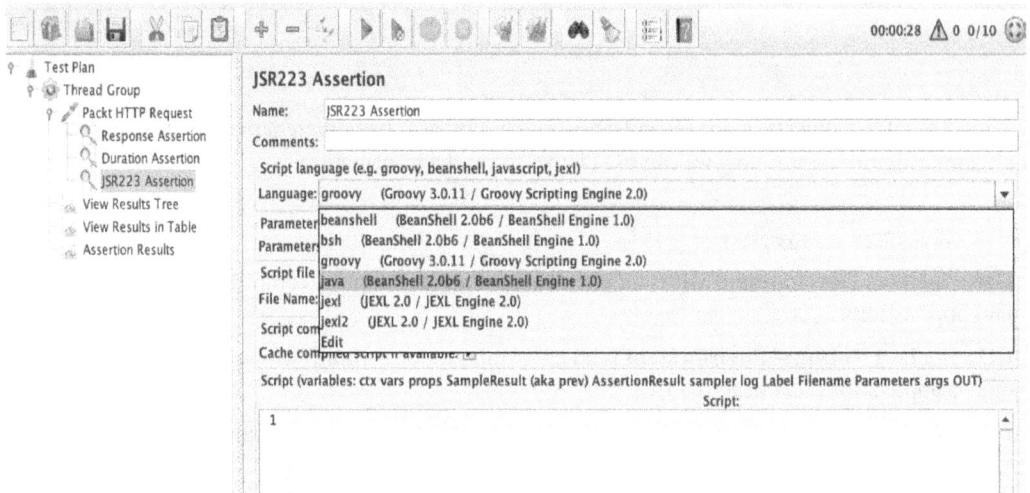

Figure 8.17 – Adding a JSR233 sampler

One of the top uses of employing custom scripting within JMeter is to enhance the logging capabilities wherever needed. This helps tremendously in reducing debugging effort. For example, the statement `log.info("Output from the log message")` can be used to print additional logging messages. Now consider the following code block, which can be run as part of the JSR233 assertion:

```
int thread_run_time = SampleResult.getTime();
int thread_latency = SampleResult.getLatency();
int response_threshold = 1000;


if ((thread_run_time+thread_latency)>response_threshold){
    AssertionResult.setFailure(true);
    AssertionResult.setFailureMessage("Threshold exceeded");
    }
```

`SampleResult` is a built-in JMeter object through which various properties of the test result can be accessed. Here, we are getting the run time and latency of the HTTP response and using an `if` statement to perform an assertion. Custom scripting thus extends JMeter's ability to perform specific validations.

Another area where custom scripting can be used is with getting and setting values of variables and properties. It might be necessary to dynamically change the value of a variable based on the test result. This can be performed with the following statements:

```
failure_count = vars.get("failure_count");

Failure_count++;

vars.put("failure_count", String.valueOf(failure_count));
```

We are getting the value of the `failure_count` variable and incrementing it before writing it out. As you can see, custom scripting opens up various ways to extend our tests to address project-specific needs. This is as far as we can go here; it's up to you to explore it further.

In the next section, let's explore some considerations for performance testing.

Considerations for performance testing

Let us quicky look at some of the major considerations required for effective performance testing:

- **Distributed load testing**: Load testing should mimic the real-world user load, and it is vital to distribute the load across multiple machines/servers to uncover critical bugs or memory leaks during the testing phase. This can be done in JMeter by employing a local JMeter master machine with remote slave machines.

- **Managing resource requirements**: The memory and resources of the load testing tool have to be managed effectively to prevent false negatives in the test results. Running through the command line, disabling memory-consuming parts of the testing ecosystem, and optimizing test data are some first areas to look into.

- **Performance testing metrics**: It is critical to review and agree upon a standard set of metrics for all the applications in your organization. This helps you standardize your test results.

- **Gradual performance degradation**: Software applications may experience a gradual decrease in throughput due to a certain code block or a faulty configuration. These issues are usually not exposed by a single load or stress test. Therefore, it is critical to run regular and prolonged load tests based on the application's needs.

- **Performance test environment**: The environment where the load tests are run should be analyzed and set up to mimic the production environment as closely as possible. Hardware, software, and network configurations should be taken into account when you are coming up with a performance test plan.

- **Constant feedback loop**: Performance testing should lead to a constant loop of performance engineering to remove bottlenecks and improve user experience. Any time a major code change is introduced, selected performance tests have to be executed and the results have to be reviewed with stakeholders.

This brings us to the end of this chapter, and it should help you quickly get up and running with your performance tests. In the next section, let's summarize what we have learned in this chapter.

Summary

We commenced this chapter by discussing what JMeter is and how it works, and then proceeded to download it. We ventured on to write our first performance test by utilizing various components within JMeter's test plan, such as thread groups and HTTP requests. We then learned how to use assertions to perform validations on the response. Subsequently, we looked at how to run the same test plan via the command line and explored using the command line to generate detailed HTML reports. We glanced through the HTTP recording capabilities of JMeter, and in the next section we dived into custom scripting by going through a quick introduction to Java and how to use a JSR233 assertion. We concluded the chapter by looking at performance testing considerations.

Questions

1.   What is JMeter and what is its primary use?
2.   How do you simulate user load in JMeter?
3.   Why is the command line preferred over JMeter's GUI for test execution?
4.   Why is custom scripting needed in JMeter?
5.   What is distributed load testing?

Part 3:
Continuous Learning

In this part, we will understand the **Continuous Integration (CI)/Continuous Delivery (CD)** methodology and how several types of testing fit into the CI/CD paradigm. We will also explore the common issues and pitfalls encountered in test automation. By the end of this part, you will emerge more confident in dealing with practical test automation matters in the current quality engineering landscape.

This part has the following chapters:

- *Chapter 9, CI/CD and Test Automation*
- *Chapter 10, Common Issues and Pitfalls*

CI/CD and Test Automation

We have been introduced to different kinds of testing so far in this book, and the chief tenet of these various test types is to produce an accurate and reliable software application at every change. This is much easier said than done. Successful testing efforts not only ensure sufficient test coverage exists but also emphasize the persistence of this coverage throughout the project life cycle. This is where CI/CD practices and techniques facilitate and improve software quality. **CI/CD** stands for **continuous integration and continuous delivery**. Test automation is at the core of CI/CD techniques, and in fact, CI/CD implementation is considered lacking without proper integration with automated tests.

In this chapter, we will be diving deep to understand the fundamentals of CI/CD and how CI/CD systems work. We will also review some test automation strategies for CI/CD and finally learn to create a job on the CI pipeline.

We will cover the following topics:

- What is CI/CD?
- Test automation strategies for CI/CD
- GitHub Actions CI/CD

Technical requirements

We will be working on GitHub Actions in the last part of this chapter to implement a CI job. The repository used will be `https://github.com/PacktPublishing/B19046_Test-Automation-Engineering-Handbook`. It is advised to possess a basic familiarity with the GitHub UI and how it works to follow along.

What is CI/CD?

CI/CD software engineering practices allow us to automate the building and testing of code to validate it meets the specifications. After all the tests have passed, they equip the teams with options to automate the delivery of code. Continuous Integration, Continuous Delivery, and Continuous

Deployment together significantly reduce the time it takes to deliver application enhancements while the engineering team can solely focus on product and code enhancements. *Figure 9.1* illustrates how these processes work in tandem to elevate software delivery levels:

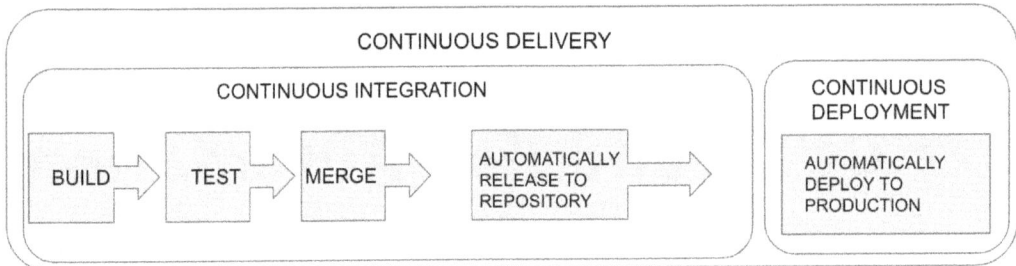

Figure 9.1 – CI/CD

Let us take a quick look at the CI/CD process in the next section.

CI/CD process

CI/CD methodologies and the associated systems take a big leap in terms of software quality improvements than any other technology has done in the recent past. Automated tests are run as part of the CI system every time a code or configuration change is made and thus serve as a comprehensive regression test bed. Modern CI systems can not only execute tests that check functional behavior but also validate the performance and security aspects of the software application. A well-implemented CI system inspires good development practices by providing constant feedback. One of the great benefits of CI systems is the visibility they provide the entire team into the release and feedback process. Any member can visualize the progress and view failures with just a click of a few buttons.

The CI/CD process starts when the developers commit their code changes into Git. The CI system picks up this change and triggers a new instance of the pipeline. At this stage, the code gets compiled, and the first set of tests (unit and component) are run. Code linting tools and code coverage analysis tools are also run at this point. If the build and tests pass, then a merge request is created by the developer to get feedback from the team. After this code review feedback cycle, the developer merges the code to the `main/master` branch, which triggers the unit/component tests again. Additionally, the code artifact created as part of the build process gets deployed to the subsequent test environment. This is where smoke and **end-to-end** (E2E) tests are run to validate business scenarios involving multiple services and/or UIs. This feedback loop continues until the code finally gets deployed to the production environment where additional tests may run.

Let us dive deep into CI in the next section and understand its nuances.

CI basics

CI addresses software integration problems effectively and efficiently. As the application under development becomes complex and the code base involves multiple components, it is vital to get feedback at every code change. The CI development practice enables engineers to integrate their work frequently and detect errors instantly through code compilation and building, and running unit tests. Each engineer usually commits at least once a day and hence encourages breaking down big changes into logical chunks of code. Building and testing of the product often occur numerous times a day in an automated manner leading to a much higher confidence level in the state of the project.

Figure 9.2 illustrates the major parts of a CI system. The CI server typically keeps polling the version control repository every minute or so and runs a preconfigured build script by fetching the latest version of the source files. The CI software has a dashboard and offers options to kick off various stages of the build process at will. The CI software is also capable of messaging the stakeholders of the current build via emails and/or instant messaging platforms. It is generally considered a good practice to perform database setup as well as part of the CI build script to ensure the coherence of the whole software application:

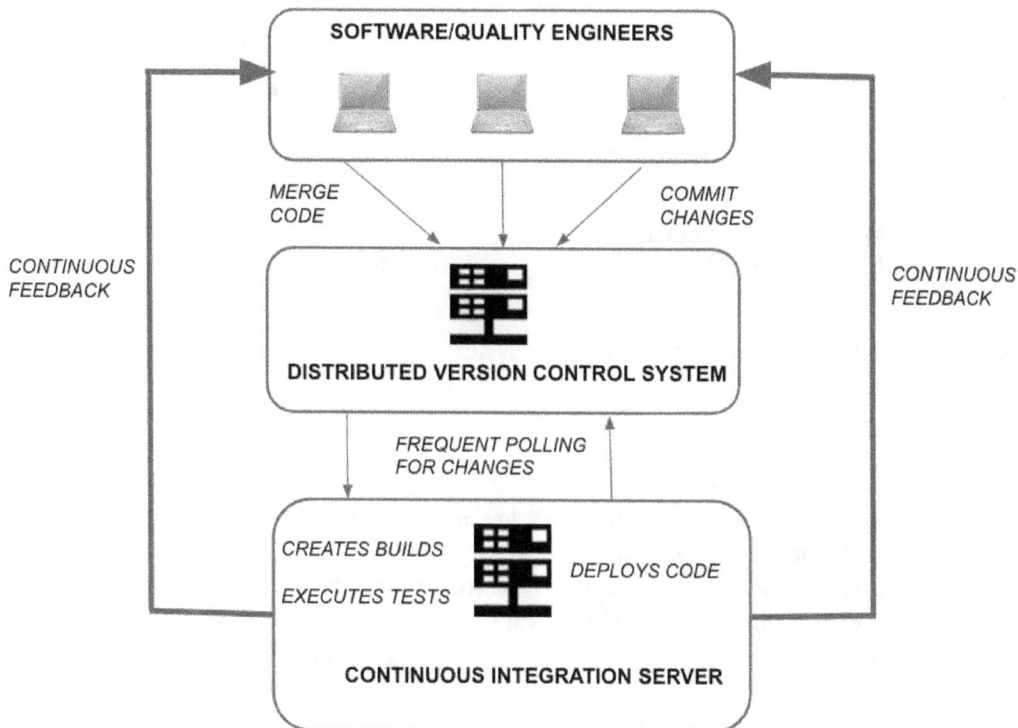

Figure 9.2 – Parts of a CI system

Some of the good development practices that result from implementing a CI system are as follows:

- Logically breaking down big code changes
- Identification of broken code faster
- Fixing broken builds swiftly
- Increased automated test coverage
- Frequent code reviews/peer feedback

CI systems efficiently enable compiling and building of source code once and deploying it to multiple environments. Even though the build's configurability may vary by platform, the CI process effectively remains the same. The usual practice is to execute the same build script against an environment-specific properties file. This makes sure we release usable software at any time and to any environment. This is a tremendous advantage and takes a lot of manual effort without CI systems in place.

Implementing a CI system in existing projects is usually a daunting task as the team may feel they have to go modify numerous processes they have been comfortable with over a period of time. It is critical to educate the team about all the extra time they gain by not performing repetitive build tasks and the overall increase in visibility from integrating individual changes into the code base.

As you may have noticed, the role of a quality engineer in the overall CI process appears minimal. But it is not so. It is vital for a quality engineer to constantly review the automated tests run as part of the pipeline to enhance coverage and reduce flakiness. This also gives quality engineers an opportunity to get involved early in the build and deployment process at every change. A quality engineer is also responsible for establishing the right quality mindset in the team. When there is a failure in the CI pipeline due to functional or E2E tests, it is often a quality engineer who responds first, debugs the failure, and logs a defect if necessary.

This brings us to the end of this section on CI. Let us next review in detail CD and the deployment pipeline.

CD and deployment pipeline

CD is the process of delivering software from development to the end user in an effective manner using a deployment pipeline. A deployment pipeline is a collection of automated processes to take the software from the engineer's machine to the end users of the product. This practice aims to curb the manual steps involved in deploying software and enables faster and more stable deployments to production-like environments. The primary aim of CD is to construct a software delivery process that is dependable and repeatable. *Figure 9.3* shows the capabilities built within a deployment pipeline. A key advantage of the CI/CD system is its ability to roll back to the most recent working version of the software when critical errors occur. Failures with configuration or code can be found quickly and reverted if necessary:

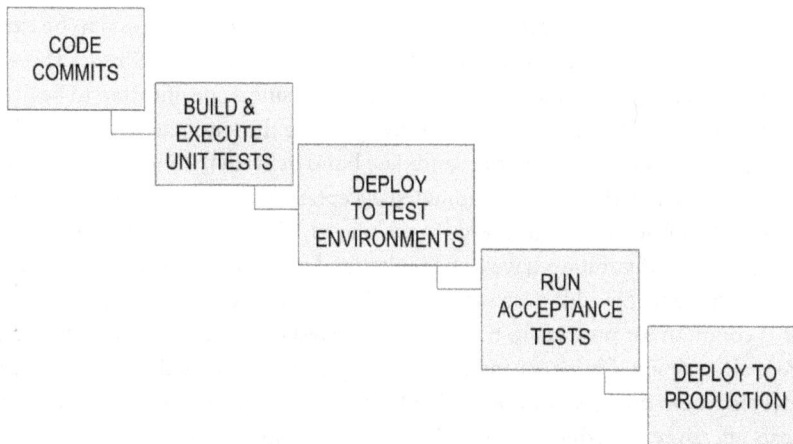

Figure 9.3 – Components of a deployment pipeline

Some of the key aspects involved in this methodology are as follows:

- Effective software configuration management
- CI
- Building and maintaining a deployment pipeline
- Automated infrastructure and environment management

Every code or configuration change introduced runs through the same set of processes, and by the time it is deployed to the production environment, we can be sure that:

- The code can be built without any errors
- The engineers are certain that their changes are working based on unit and functional tests passing as part of the pipeline
- Product and business teams accept the change due to the successful run of E2E and user acceptance test cases as part of the pipeline
- The delivery ecosystem is in place because the code was deployed and tested in a production-like environment
- A copy of the latest code and configuration change exists on the version control repository

As appealing as it may sound from the outside, implementing a CD process is daunting and takes the effort of many teams working in unison. The key lies in examining and understanding the existing deployment system, breaking it down into simple and repeatable steps, and using the latest technologies to automate it. It can be done incrementally so that over time, it evolves into a refined and efficient system.

One of the central aspects of the CD process is to automate the acceptance tests to be executed as part of the deployment pipeline. These acceptance tests may be E2E UI or API tests. These have the significant advantage of blocking a build that has failed a key business specification to be deployed to production or even a production-like environment. By updating the automated tests consistently for every code change, the tests ascertain that the candidate build delivers the intended value to the end users. A common drawback with running automated acceptance tests as part of the pipeline is that they take too long to complete for a single merge to the master. This is mainly because the underlying application and all of its configuration have to be built and deployed before any of the pre-requisite steps for the tests can be run. This adds time to the overall acceptance test cycle when the test run times are included. It is common for projects to have acceptance tests that run a few minutes to multiple hours, but there is always room for increasing the efficiency of the automated tests. Quality engineers and **software development engineers in test** (**SDETs**) should always be on the lookout for ways to refactor their tests effectively to reduce the overall deployment pipeline runtimes.

Another major advantage of the CD process is its capacity to validate certain **non-functional requirements** (**NFRs**) as part of the deployment pipeline. It is not uncommon to unearth architectural discrepancies in the late stages of a project life cycle, the reason being the nature of the NFRs themselves and also the availability of all the components to execute sensible tests against them. This challenge can be addressed through CD by including tests that validate certain performance thresholds for the application as part of the CI pipeline. It is not possible to design an extremely performant and scalable application at the inception of a project. But having these tests provides an initial comfort level with a code change, and if there are considerable performance impacts, further load tests can be performed in a full-fledged performance environment to identify the bottleneck.

There should be additional considerations made in terms of security and data compliance when working with cloud platforms. It is a good practice to include cloud network connectivity and security tests that run at regular intervals as part of the CD process. This ensures that the components are all connected and working together in a holistic cloud environment in a secure manner.

CD involves not just the collaboration between engineers and the infrastructure team but demands support from product owners/managers, executive sponsors, and everyone involved in between. It is a paradigm shift and the new norm in the software delivery world. In the next section, let us take up some considerations for test automation in the world of CI/CD.

Test automation strategies for CI/CD

Having understood what a CI/CD system entails, let us now review a few test automation strategy considerations for CI/CD. It is much easier to build test automation into CI systems earlier in the project rather than adding it later on. It is essential to understand that all tests cannot be run all the time since that would be an enormous overhead on the CI pipeline as the code base increases. Having the right test automation strategy to utilize CI/CD systems frees up engineers' capacity that can be utilized in other areas. It goes without saying that building an accurate and reliable software application is possible only by establishing quality right from the lowest building block. Let us start by looking at unit and component tests.

Unit/component tests

Unit tests verify the behavior of the smallest blocks of code, usually a single class or an object, whereas component tests exercise larger blocks of code involving multiple classes/objects. Both these tests have minimal external dependencies and often involve mocking of objects that are outside the validation logic of the test. They are usually the fastest to run and hence provide the quickest feedback. All unit and component tests should be run as part of the CI pipeline, and the recommendation is to run them on every commit to the branch and on every merge to the `master` branch. This paves the way for quick and efficient debugging when a failure occurs due to change. It also gives engineers the much-needed confidence that their changes work in tandem with other components upon merging the code. Let us next take a look at API tests.

API tests

API tests, as we have seen earlier, verify the functional correctness of API endpoints. These tests focus mainly on the business logic of the API as opposed to unit tests, which deal with much smaller units of code. API tests should be automated and run as part of the CI pipeline as much as possible on every merge to the `master` branch. Since the advent of microservices, it has become a common practice to break down business logic into multiple microservices, each containing a collection of API endpoints. Each microservice essentially has associated API tests that should be run on every code merge into the repository. This lays a perfect foundation for a wider test coverage for the core business logic and running these as part of the CI pipeline rapidly flags alterations to the existing business logic.

Running API tests for a specific microservice in a CI system is a ripe ground for using containers as these tests do not demand the presence of all dependent services. Dependent services, if any, can be mocked to ensure the continuity of the tests.

Next, let us take up how E2E tests can fit into CI pipelines.

E2E tests (API and UI)

E2E API and UI tests form a critical part of the feedback on a CI pipeline. These are usually the longest-running tests as they exercise the complete system under test and also due to the additional setup they require. They require a fully installed software application, including the databases in a test or staging environment, with all the dependencies in place. Vendor service calls may be mocked in these tests to save costs. These longer-running tests can be triggered by the CI system and run in an isolated test environment at planned intervals. They can also be invoked by the CI pipeline as part of a job that deploys application code to a specific environment.

UI tests in general tend to be brittle when compared to API tests, and it is essential to hand-pick a selection of E2E tests to run in the context of CI. Long-running CI pipelines can damage an engineering team's productivity, and hence it is crucial to include only stable tests. Running these tests on every commit or build could spell disaster, and it is mandatory to monitor and optimize the CI pipeline's runtimes whenever an E2E test case is added to the test suite.

Let us next look at smoke tests and how they can be utilized in a CI system.

Smoke tests

Smoke tests do not address a specific architectural component; instead, they focus on delivering a clean build. Smoke tests are a collection of test cases that are run at the end of a deployment job to give confidence in the quality of the build. They are selected based on the business use cases and workflows in the application and verify they work as per the specifications on every deployment to an environment. Smoke tests are a combination of E2E API/UI and component tests. Though minimal and conservative, there are smoke tests that can be run against the production environment. Major CI/CD systems and test automation frameworks provide features such as tagging that help flag smoke tests with ease. Additional precautions have to be taken to organize environment-specific variables in the CI/CD system if the same smoke test suite is designed to run against multiple environments.

Table 9.1 summarizes the content discussed in this section:

Type of Test	Recommended CI/CD Strategy
Unit/component tests	Tests with minimal dependencies and the quickest feedback cycle to be run on every commit and every merge to master
API tests	Tests that verify the functional correctness of the API endpoints to be run on every merge to master
E2E API tests	Long-running tests involving sequential API calls to test business workflows to be run on every deployment to test environments
E2E UI tests	Long-running tests involving user actions to test business workflows to be run on every deployment to test environments
Smoke tests	A subset of tests selected to be run on every deployment to an environment

Table 9.1 – CI/CD strategies for various test types

Figure 9.4 illustrates how a CI/CD system incorporates constant feedback into the development process:

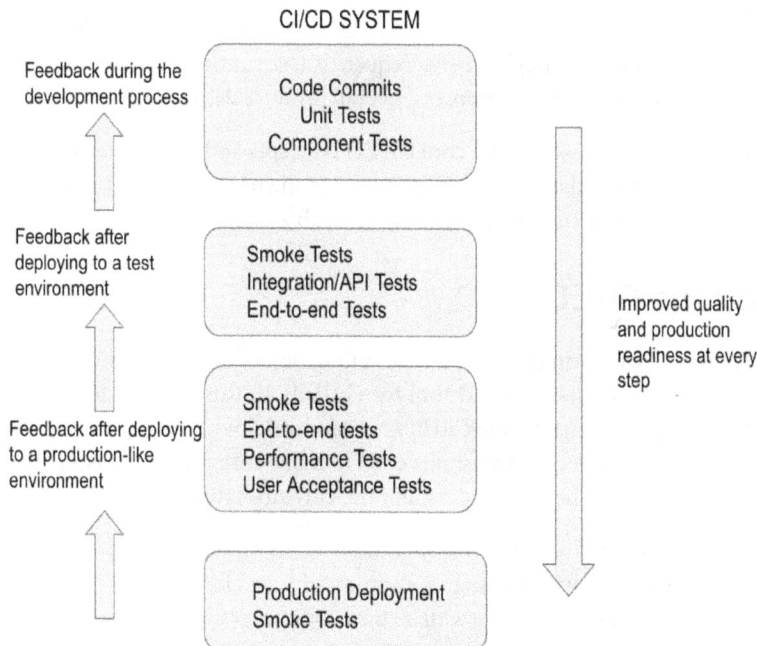

CI/CD SYSTEM

Feedback during the
development process

Code Commits
Unit Tests
Component Tests

Feedback after
deploying to a test
environment

Smoke Tests
Integration/API Tests
End-to-end Tests

Improved quality
and production
readiness at every
step

Feedback after deploying
to a production-like
environment

Smoke Tests
End-to-end tests
Performance Tests
User Acceptance Tests

Production Deployment
Smoke Tests

Figure 9.4 – Feedback loop in a CI/CD system

Next, let us review some tips on how to address test failures in a CI pipeline.

Addressing test failures

Persistent test failures can occur within a well-organized CI pipeline. It is critical to identify and address the root cause as quickly as possible. Consistent test failures render the CI/CD process ineffective, and hence measures should be taken to improve the success rates of the pipeline.

Some common causes of the failure of a pipeline are as follows:

- Interactions between the test framework and the CI system are not well defined
- The test suite running on the pipeline is too big
- Insufficient test reporting
- Lack of collaboration between quality engineers, software engineers, and infrastructure teams

Quality engineers should examine the pipeline frequently and always have a backlog of improvement items lined up. Optimizations to the CI pipeline from a test automation perspective can be made by consistent refactoring of tests to reduce flakiness. Long-running tests should be set up for parallel execution wherever possible. There can be instances where the same tests are being run at the same time in multiple pipelines. Data setup and teardown steps for each test should be built into the test to account for such cases. CI/CD systems store a lot of pipeline-related data that can be used to generate reliable metrics such as production deployment frequency, test runtime, test failure rate, and so on. These metrics provide helpful insights to improve overall product delivery times.

So far in this chapter, we have looked at the core CI/CD concepts and theoretically understood how test automation can be accommodated in our CI pipeline. In the next section, let us acquire hands-on experience of implementing one such pipeline using GitHub Actions.

GitHub Actions CI/CD

GitHub Actions is a CI/CD platform that enables the automation of building, testing, and deployment of application code. It is the built-in CI/CD tool for GitHub. In this section, let us go over all the concepts we need to know to understand the GitHub Actions workflow. We will also learn to implement a GitHub action to run syntax checks against our code to make sure it meets specific criteria. Let us start with the necessary terms to help us understand the GitHub Actions workflow file.

The workflow .yaml file contains all the information used to initiate and drive the CI pipeline to completion. YAML is a data-serialization language commonly used for building configuration files. It is in human-readable format and compatible with all the major programming languages. The workflow .yaml file at a high level specifies the following:

- **Events**: An event is a trigger for a workflow
- **Jobs**: Jobs are high-level actions performed as part of the workflow
- **Runners**: A runner is a platform where the action is performed
- **Steps**: A job can be broken down into multiple steps
- **Actions**: Each step performs a specific action in an automated fashion

For illustration, we will be using code commits and merges, which are common events that occur in every repository. In this example, we will be configuring our workflow file to be triggered when someone pushes code to our repository. When this push event occurs, all jobs within the workflow will be run. This is demonstrated by the YAML code snippet shown next. In this configuration file, we use the on parameter to specify the trigger for the workflow. When the push event occurs, it will run all jobs within this workflow. We have a single job here that comprises multiple steps and actions. Under the steps, two actions will be run in this case. The first action will check out the latest version of our code from the main/master branch, and the next one will run the super-linter against it. Linters are tools to evaluate that our code conforms to certain standards. The super-linter supports multiple languages and automatically understands and checks any code in the specified repository. The runs-on parameter is used to specify the runner. This is the container environment where

GitHub will run this job. There are additional options to host your own container; however, we will be sticking to the default container offered by GitHub in this case:

```
name: Packt CICD Linter Demo

on: [push]

jobs:
  super-lint:
    name: Packt CICD Lint Job
    runs-on: ubuntu-latest
    steps:
      - name: Checkout Code
        uses: actions/checkout@v3

      - name: Lint Code Base
        uses: github/super-linter@v4
        env:
          DEFAULT_BRANCH: main
          GITHUB_TOKEN: ${{ secrets.GITHUB_TOKEN }}
```

Let us now go to GitHub to set up a workflow in our repository (https://github.com/PacktPublishing/B19046_Test-Automation-Engineering-Handbook). First, we create the right folder structure for our workflow file. We use the **Add File** option on the home page of our repository. We create a linter_demo.yml file with a .github/workflows structure under the root folder of the project and copy the code into the editor below, as shown in *Figure 9.5*. Then, this file can be committed through a new branch or to the main branch directly. It is mandatory to follow this folder structure to save the workflow file:

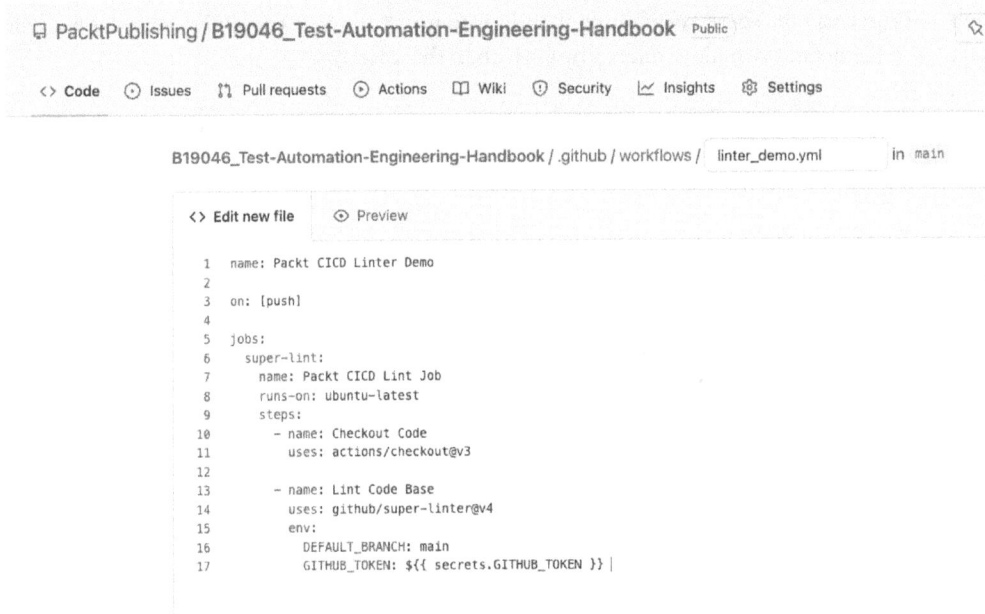

Figure 9.5 – Creating a GitHub workflow file

On navigating back to the home page of the repository, we notice a yellow status icon now, as shown in *Figure 9.6*. This signifies that the workflow is being run now and the code is being checked. This status icon turns green or red based on whether the checks pass or fail. This is particularly helpful when you are viewing a new repository and it aids to know that the repository is in a healthy state with all the tests passing. The results of the workflow can be viewed by clicking on the status icon or visiting the **Actions** tab:

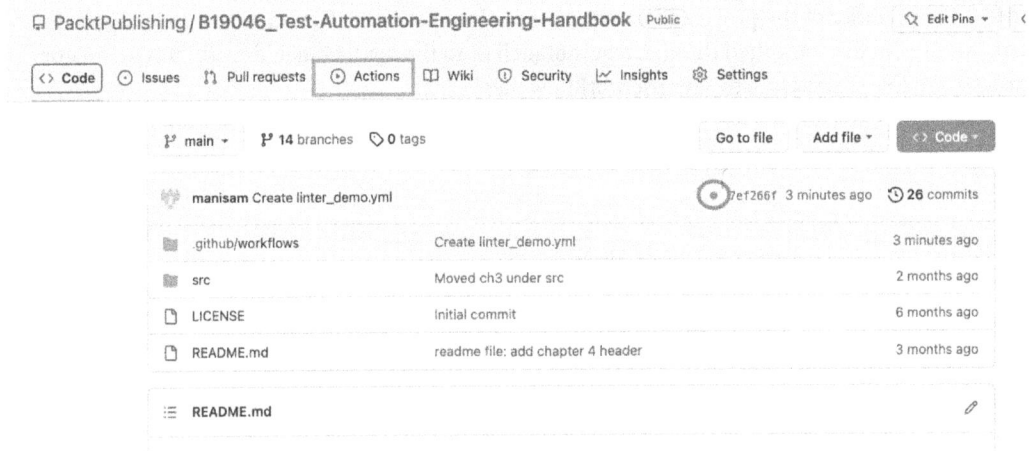

Figure 9.6 – Workflow status

We can view the execution results of a specific job by following the link within the **Actions** tab. This provides a neat breakdown of the steps executed within the job and how long each one took. You could open each step to view the run logs. *Figure 9.7* shows the view for a failed job and its individual steps executed as part of the workflow:

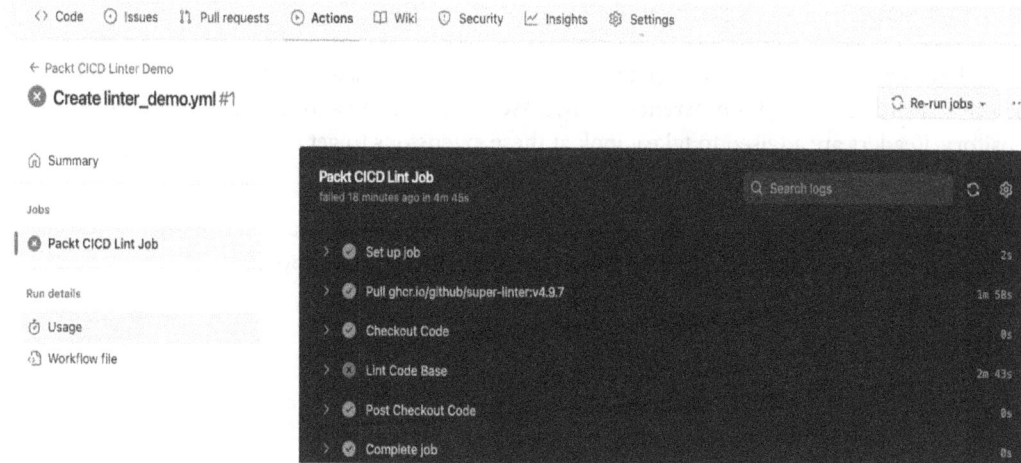

Figure 9.7 – Workflow results

The **Actions** tab is where all the CI/CD information is shown within a GitHub repository. It shows a history of all our workflow jobs and their statuses, with options to look through each one further in detail. We can have as many workflows as we need within a single hub repository. For example, we could have one workflow that runs only Cypress tests and another to lint the entire code repository.

On fixing the suggestions from the linter and pushing the code to the repository, the CI job should automatically be triggered based on our setting in our workflow file.

The following is a sample snippet to invoke Cypress tests for reference. Placing these contents in a workflow file at the root of the project under the recommended directory structure triggers Cypress tests on every commit to the repository:

```
name: Packt Cypress Tests
on: [push]
jobs:
cypress-run:
name: Packt Cypress CI/CD Demo
runs-on: ubuntu-latest
steps:
- name: Checkout Code
    uses: actions/checkout@v3
```

```
 - name: Cypress.io
    uses: cypress-io/github-action@v4.2.0
env:
DEFAULT_BRANCH: main
GITHUB_TOKEN: ${{ secrets.GITHUB_TOKEN }}
```

GitHub has an extensive marketplace (https://github.com/marketplace) where you can grab pre-written workflows for different use cases. We can download and modify them to use in our repository. Readers are advised to take a look at these extensions to get an idea of the tremendous community surrounding CI/CD systems.

This brings us to the end of this chapter. In the next section, let us quickly summarize what we learned in this chapter and peek into our explorations in the final chapter of this book.

Summary

We commenced this chapter by understanding what a CI/CD system entails and the processes involved. We then dived deep into CI and CD individually to achieve a better understanding. Then, we took up test automation strategy considerations with respect to CI/CD processes. In the last section, we did a hands-on exercise to create a CI job using the GitHub Actions tool. In the next chapter, we will be learning about the common issues and pitfalls when working with test automation.

Questions

1. What is CI/CD and why is it necessary?
2. How is a CI process triggered?
3. What does a CI server do?
4. What are the aspects involved in CD?
5. How often should E2E tests be run in the CI pipeline?
6. What is GitHub Actions and what is it used for?

10
Common Issues and Pitfalls

Test automation, with all its benefits, helps engineering teams save time, effort, and resources. It takes highly collaborative and skilled engineers to get test automation working at a large scale. Even high-performing teams tend to go through multiple iterations before settling on a stable framework. Teams typically encounter a wide variety of issues when it comes to test automation, and each team's journey is unique. In this chapter, I have tried to compile a list of teachings to guide you through this process and help minimize any hurdles in your test automation undertaking. The main topics we will be covering in this chapter are the following:

- Recurrent issues in test automation
- Test automation anti-patterns

Recurrent issues in test automation

Being successful in test automation involves getting a lot of things right – in this section, we will look at a collection of common issues encountered when executing test automation projects. These items throw light on common viewpoints for anyone undertaking test automation at a large scale. This is not meant to be a comprehensive list, but these are more common occurrences that I have witnessed over time.

Unrealistic expectations of automated testing

Engineering teams come under pressure quite often to meet unrealistic expectations from automated tests. It is the duty of SDETs and quality engineers to educate the stakeholders about the initial investment it takes to see results from test automation. Going all in on a framework without understanding what it can or cannot do for your organization often leads to wasted resources. Another common issue is stakeholders questioning the need for multiple frameworks to address the application stack. This also demands resources that are readily available and skilled with that knowledge. The initial leg of the test automation journey is overly critical and needs strong collaboration between management and engineering teams.

Let us now look at the importance of manual testing before kicking off automation.

Inadequate manual testing

Good test automation coverage usually springs from extensive manual testing. Every feature included as part of a test automation suite should be thoroughly manually tested and considered stable enough for automation. Manual testing gives us the essential confidence to move to test automation. In fact, the manual testing of a feature is where most of the bugs are found, and automation is an ally for catching regression bugs at a later point. Manual testing should not just be limited to functionality but also help find issues with application configuration and test environments. Debugging a test automation script becomes trickier when there are multiple layers of failure. Another important aspect of manual testing is that it produces a vast sample of test cases to choose from for test automation. Therefore, early in the development cycle, test case creation should be completed, and these test cases can be executed against a stable build. At this point, the foundational work of test automation can begin, and when we know that the feature is fully working later, the flow can be automated and added to the corresponding test suite.

Let us review the right candidates for test automation next.

Not focusing on automating the right things

Just because we have a framework in place and the resources lined up for automation, we should automate everything. It is imperative to avoid going down the path of 100% test automation. A lot of consideration should go into calling out the right candidates for automation. Some of the key items to consider are as follows:

- Pay special attention when automating the tests that do not need to be run frequently. Automate rare cases only when absolutely necessary.

- Automation costs are initially high and we will only start seeing the ROI after several rounds of execution, so it is all-important to focus on stable business scenarios that can reap the benefits in the long run.

- Minimize automating tests at the UI layer due to the brittleness of the frontend. Load the base of the test automation pyramid as much as possible.

- Do not automate the usability aspects of your application.

- Avoid automating big and complex scenarios in the early phases of an automation framework. Automating a complex scenario in a single flow will impact test stability and eventually result in higher maintenance costs if the framework is not mature enough to handle it.

- Brittle selectors are the biggest cause of flakiness in the UI tests. Working with your team and identifying a solid selector strategy saves a lot of time in the long run.

Test engineers should strive to only automate the required and right set of test cases and should push back if anyone says otherwise.

Let us now look at how important the architecture of the underlying system is to test automation.

A lack of understanding of the system under test

Every software application is unique and so must be the test suite designed to validate its features. Test engineers often struggle to automate features reliably due to their lack of a deeper understanding of the various aspects of the application. Test engineers should collaborate with software engineers at every stage of the development process. There can be intermittent UI test failures due to subtle backend changes and test engineers should strive to say on top of the code changes that are made. A single API call or a web page can no longer be viewed as a black box, as it may add various asynchronous activities behind the scenes down the road. If the test suite is not ready for these changes, a lot of refactoring work might be needed to get it operational again. Therefore, it is mandatory for engineers to take the time to get an in-depth understanding of the system and then go ahead with designing the automated tests.

Let us understand the importance of test maintenance next.

Overlooking test maintenance

One of the easiest aspects to overlook in test automation is its maintenance. A lot of teams try to ramp up test automation infrastructure when product delivery is in full swing and they start witnessing regression defects leaking into production. They just want to be able to get up and running as quickly as possible with a test automation infrastructure without giving much thought to the maintenance aspects. This is a recipe for disaster, as it is a known fact that any design work that doesn't consider maintenance aspects doesn't scale very well. In the early stages of a software project, it is vital to think through what kind of effort will need to go into maintaining a test automation framework after it is fully built and functional. This usually includes items such as test data refreshing, keeping test libraries updated, addressing major architectural changes, and so on.

Let us now understand how important the right tools are for automation work.

Not choosing the right tools

The test automation market is flooded with a wide variety of tools, both licensed and open source, and it is essential to pick the one that suits the needs of your organization. Not selecting a tool just for its popularity and thinking about all the expectations from the tool practically saves a lot of effort down the road. Test coverage-related lapses often stem from not analyzing how the selected tool addresses the entire application stack. It is essential to examine how the test automation tool provides coverage for every layer of the test pyramid. Not all the tools on the market come with all the capabilities needed and most often, it is up to the end users to customize their choices to fit their needs.

Let's review the impact of insufficient investment in test environments next.

Under-investing in test environments

A test environment is arguably the key factor that decides the success of a test automation endeavor. Not investing properly in the test environment and expecting automated tests to be stable is usually wishful thinking. Based on the type of automated tests, the test environment should have all the necessary pieces installed and ready to use. For example, the environment used for end-to-end testing should be close to production in all aspects, except for load handling. The current infrastructure landscape supplies tools such as **Garden.io**, which can be used to spin production-like environments at will for local development and testing. Software engineers should have the flexibility to bring the test environment as close as possible to their code rather than testing first in a full-fledged environment.

Another critical failure point concerning test environments is not considering how they work with a **Continuous Integration** (**CI**) tool. Automated tests are usually packaged and deployed on the CI platform to run tests against an environment. It is vital for these three components to work together without any incompatibility.

Let us look at the value of taking a whole team approach to test automation next.

Taking a siloed approach

A successful test automation approach often involves the collaboration of the entire team. Keeping a quality-first mindset brings different skill sets and viewpoints to the table and increases the testability of the product. Having the whole team approach to test automation problems results in higher test coverage, efficient testing processes, and, eventually, higher confidence in the product released. Taking a whole team approach to test automation forces every engineer to think about how to test a piece of code, thereby evolving quality as the product is being built rather than as an afterthought. It also helps with better collaboration and ownership.

Let us now look at how being lean with automation helps overall.

Not taking a lean approach

A general recommendation when it comes to test automation efforts would be to not work on any large-scale changes for more than two iterations. It is important to keep trying something new, but it is also wise to take the time to appraise what is going right and what is not working every iteration or two. This gives us the flexibility to change direction when things are not going right or accelerate when the results are encouraging. The last thing we need is to spend too much time on a tool or a framework and find it sub-optimal. There is no one-size-fits-all solutions when it comes to test automation, and it is crucial to keep the approach lean.

Let us next realize the importance of having a plan for test data.

Not having a plan for test data needs

Sound test automation strategies devote ample time to collecting, sanitizing, and managing test data. Test data is at the core of most test automation frameworks. Ignoring test data needs and only focusing on other technical aspects often results in a framework that cannot be scaled. I would go one step further and say that test data is as important as the framework itself. Having a plan for data seeding at every layer of the application goes a long way in building robust automated tests. It is equally important to acknowledge the areas where data is hard to simulate and, in turn, look for alternatives such as mocking or stubbing techniques. Test results are deterministic only when the input data is reliable and predictable.

Frequent baselining production data and migrating it to test environments helps build a reliable test data bed on which automated tests can be executed. Automated tests should also be built in a self-contained manner when it comes to handling test data. Proper clean-up steps should be built within the frameworks to account for test stability. More often than not, proper test coverage equates to having sensible data to drive automated tests.

This brings us to the end of this section. In the next sections, let us review certain commonly occurring anti-patterns in the world of test automation.

Test automation anti-patterns

An anti-pattern is something that goes against the grain of what should be done in a project. It is important to be mindful of these so that you may recognize and avoid the occurrence. These anti-patterns can be coding- or design- as well as process-oriented. Let us begin with a few coding- and design-based anti-patterns in the next section.

Coding and design anti-patterns in test automation

Test automation code should maintain extremely lofty standards when it comes to stability and reliability. Poorly designed tests result in more time spent on debugging tests rather than developing new product features. This has a ripple effect across teams, thereby leading to the overall degradation of quality and productivity. It is essential to focus on the basics and get them right from the bottom up. Let us begin with the test code quality.

Compromising the code quality in tests

There should be no differentiation between tests and core application logic when it comes to code quality. Test code must be treated the same way as any other code would be in each repository. Applying coding principles such as DRY, SOLID, and so on to test code helps keep the standards up. Test engineers are often tasked with getting the team up and running quickly with an automation framework. Even though this is important, immediate follow-up tasks should be created to make the framework more extensible and maintainable. Eventually, most of the test framework maintenance work will end up in the backlog of test engineers and they should think through the code quality at every step.

Let us now understand the effects of tightly coupled code.

Tightly coupled test code

Coupling and cohesion are principal facets of coding that should be paid special attention to when it comes to test code. Coupling refers to how a particular class is associated with other classes and what happens when a change must be made to a specific class. High coupling results in a lot of refactoring due to strong relationships between classes. On the other hand, loose coupling promotes scalability and maintainability in test automation frameworks.

Cohesion refers to how close the various components of a class are knit together. If the method implementation within a class is spread out across multiple classes, it gets harder to refactor, resulting in high code maintenance efforts. For example, let us say we have a class that contains all the methods to assert various responses from a single API endpoint. It would be termed loose cohesion if you added a utility within the same class to generate reports from these assertions. It might seem related at a high level, but a reporting utility belongs in its own class, and it can be reused for other API endpoints as well.

Let us next review the impact of code duplication.

Code duplication

Code duplication happens in a lot of test frameworks, most often to get the job done quickly rather than taking time to design it the right way. This hurts the maintainability of automation frameworks eventually. It is essential to refactor and optimize test code frequently. *Figure 10.1* illustrates a simplified case of code duplication where there are two classes to generate distinct kinds of reports. In this case, methodC and methodD are the only differences between the classes. The rest of the class body is duplicated, forcing multiple tests to instantiate both these classes every time they need to generate reports. This results in a maintenance nightmare when there is a change in the way reports are generated or when these classes are combined.

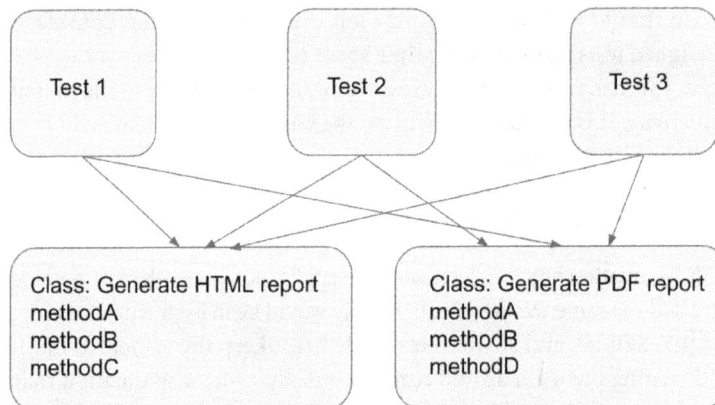

Figure 10.1 – Code duplication

It is important to remember the *rule of three* when it comes to code duplication. Whenever there is a need to duplicate a chunk of code a third time, it needs to be abstracted into a separate method or class.

Let us now look at how lengthy and complex tests affect test maintainability.

Lengthy complex tests

Extending existing tests and making them more complex often results in brittleness and fragility in the test suite. Engineers tend to add to the existing tests as the project scope increases and more features are being delivered. This may seem like a viable solution in the short term, but it results in complicated code for the rest of the team to follow. We have a code snippet here that shows a long test that navigates to four different screens and validates multiple fields in the process. This can be easily broken down into single tests in their own blocks or spec files if necessary:

```
describe("Visit packt home page, ", () => {
beforeEach(() => {
cy.visit("https://www.packtpub.com");
});
it("search, terms and contact pages", () => {
const search_string = "quality";
const result_string = "Filter Results ";
cy.get('#__BVID__324').and("have.value", "");
cy.get('#__BVID__324').type(`${search_string}`,
{        delay: 500 });
cy.get('.form-inline > .btn-parent > .btn >        .fa').
click();
cy.get(".filter-results").contains(result_string);
cy.get(".reset-button", { timeout: 10000 }        ).should("be.
disabled");
cy.get("#packt-navbar").and("have.class", "navbar-    logout");

//test term-conditions page
cy.visit("https://www.packtpub.com/terms-          conditions");
cy.get('.form-inline > .btn-parent > .btn > .fa').click();
cy.get(".terms-button", { timeout: 10000 }).should("be.
enabled");

//test contact us page
cy.visit("https://www.packtpub.com/contact");
```

```
cy.get('.form-inline > .btn-parent > .btn > .fa').click();
cy.get(".send-button", { timeout: 10000 }).should("be.
disabled");
});
```

Long and complex tests eventually result in adding timeouts in multiple places to account for slowness. This results in increased flakiness and long-running test suites. Therefore, it is important to break down the test code to follow the single responsibility principle as much as possible. It is hard to achieve this, especially in end-to-end tests, where there is usually a business flow involved, but constant efforts must be made to keep it as close to a single responsibility as possible.

Let us next review the ways to employ assertions.

The incorrect use of assertions

Assertions are our primary instruments for validating the application logic in a test. They should be used correctly, sparingly, and sensibly. One surprising area of concern is when assertions are not used at all. Engineers tend to use conditionals or write to the console instead of employing assertions, and this should be avoided at all costs. Another pain point is when there are multiple assertions within a single function or a test. This negates the primary logic of the test , thereby leading to undesirable results. A third way in which assertions can be misused is when the right type of assertions are not employed.

The following code snippet illustrates a case where no assertions are being used. This is strongly discouraged in a test:

```
function compute_product() {
...Test logic...
    if (product==10){
        console.log('product is 10');
    else if (product==20){
        console.log('product is 20');
        }
    else {
        console.log('product is unknown');
        }
    return product;
}
```

The next code snippet illustrates the use of multiple assertions in a single test. Use them sparingly based on the test type and scenario:

```
cy.get('[data-testid="user-name"]').should('have.length', 7)
cy.get('[data-testid="bank_name"]').should('have.text', 'BOA
Bank')
cy.get('[data-testid="form_checkbox"]')
  .should('be.enabled')
  .and('not.be.disabled')
```

Lastly, the following code snippet shows the use of the same type of assertions for multiple UI elements. Explore and employ the appropriate type of assertion based on the element being validated:

```
cy.get('#about').contains('About')
cy.get('.terms')contains('terms-conditions')
cy.get('#home').contains('Home')
```

Let us look at data handling within test frameworks next.

Mishandling data in automation

Data can be mishandled easily – sometimes with dire consequences. There are different types of data that a test automation framework handles and stores. Each type of data has its place within the framework and it should be handled as such. Mixing them up will result in an unorganized and inextensible framework. Some of the common types are listed with examples:

- **Functional test data**: This drives the application logic and is seeded within the framework or comes from a test environment.

- **Dynamic test suite data**: This is data required by the test scripts for execution, such as secrets:

```
node test-script.js -secret='HAGSDH' -timeout=30000
```

- **Global data**: This is configuration data specific to particular environments, stored in `config` files and the CI system:

```
DEV_URL= //test-development.com
STAGING_URL=https://test-staging.com
AWS_KEY=test-aws-key
```

- **Framework level constants**: These are constant values required by the tests and stored within the framework in a non-extendable base class:

```
const swift_code = 111222333,
      routing_number = 897654321;
class BankConstants {

  static get swift_code () {
    return swift_code;
  }

  static get routing_number () {
    return routing_number;
  }
}
```

Let us now understand the importance of refactoring.

Not refactoring according to changing requirements

It is necessary to refactor test code according to changing requirements. Unfortunately, there are times when additional requirements get added or existing requirements are modified beyond the scope of an iteration. If the framework is not flexible enough to accommodate refactoring, it is extremely difficult to keep up with these changes. In an Agile world, it should be a top priority to make the test automation framework extensible so that these changes can be made.

Let us review the use of UI tests to validate business logic next.

Validating business rules through the UI

UI test automation should be restricted to testing high-level business flows, and all the business rule validations should happen at the API level as much as possible. Many teams have wasted their valuable resources by using UI automation to cover low-level business rules due to scripts breaking at minor code changes. The initial overhead is way too high to realize any ROI with this approach. It is extremely hard to acquire fine control over the system under test through UI components. These are also the slowest running tests in the test automation pyramid and hence deprive the teams of valuable rapid feedback.

Let us now look at the effects of inefficient code reviews.

Inefficient code reviews

Code reviews, when not done correctly, lead to code that does not meet the quality standards of an organization. Test automation code reviews are usually not performed at the same level as the code delivered to customers. An inefficient code review process may lead to a lack of code quality standards and the introduction of new bugs into the existing framework. Chief items to look for in a test automation code review are as follows:

- Proper commenting and descriptions wherever necessary
- Formatting and typos
- Test execution failures
- The hardcoded and static handling of variables
- Potential for test flakiness
- Design pattern deviations
- Code duplication
- Any invalid structural changes to the framework
- The correct handling of `config` files and their values

That brings us to the end of our exploration of coding- and design-oriented anti-patterns. Let us now explore a few process-oriented anti-patterns in test automation.

Process-oriented anti-patterns in test automation

Building a test automation suite and maintaining it operationally is no easy task. Apart from the technical obstacles, there are a variety of process-related constraints that hinder efforts. In this section, let us review a few process-oriented test automation anti-patterns.

Test automation efforts not being assessed

It is a widespread practice for software engineers to assess the effort required to complete the work of building a feature. In the Agile landscape, this is usually represented in story points. However, this pattern unfortunately is broken quite often for test automation, resulting in the accumulation of technical debt. Every feature or story being worked on should be evaluated the same way for test automation as it is done for manual testing and release. It is a common occurrence in the industry to punt the test automation efforts to a later iteration due to tight deadlines and other technical limitations. This can simply be avoided by considering test automation as an integral part of completing a feature. A feature should not be considered complete unless the team feels they have sufficient automated test coverage for it.

Table 10.1 shows a sample matrix that can be used to estimate test automation efforts in an Agile setup. It can be customized to suit the team's needs and skill sets:

Story points	Test type	The complexity of the task	Dependencies	The effort required in days
1	API integration	Very minor	Nothing	Less than 3 hours
2	API integration	Simple	Some	Half a day
3	UI end-to-end	Medium	Some	Up to 2 days
5	UI end-to-end	Difficult	More than a few	3 to 5 days
Split into smaller tasks	API/UI	Very complex	Unknown	More than a week

Table 10.1 – Sample test automation effort matrix

Test engineers should be vocal about this aspect, as it takes time, effort, and commitment from the whole team to get quality right. It is a process, and it evolves over time to yield fruitful results.

Let us understand the impact of starting automation efforts late next.

Test automation efforts commencing late

It is a frequent occurrence even in today's Agile landscape that test automation doesn't start until the development of a feature is fully complete. Test engineers wait for the feature to be completed and then begin planning for automation. This eventually demands a big push from the team to get tests automated and, in turn, affects the quality of the output. It is true that there might be a few unknowns in terms of the exact implementation details when the feature is being developed, but the ancillary tasks required for test automation can be started and some may even be completed while the development is in progress. *Figure 10.2* illustrates how various test automation tasks can fit into a development cycle.

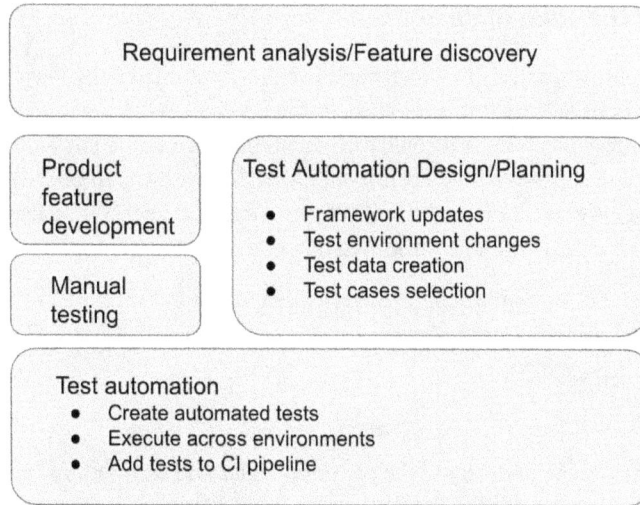

Figure 10.2 – Test automation tasks

Often, teams find it really hard to compress all the test automation activities post-feature completion, which results in project delays or, even worse, sub-par product quality by leaking defects into production.

Let us look at how proper tracking helps with test automation next.

Infrequent test automation tracking

It is a common project management practice to track as much information as possible in an ongoing project, but this pattern gets broken in the Agile landscape, where there is a lot of emphasis on face-to-face communication rather than documentation. Engineers spend all their time in the code bases, not devoting time to tracking what has been done so far. Certain basic aspects must be documented for several reasons. For example, teams might be juggled due to an organization being restructured and new members need to know where to start. Another common case in which the team suffers is when an experienced test engineer leaves the organization. Some critical items to keep a track of when it comes to test automation are as follows:

- Documenting what is automated and not automated

- Noting which tests are not worth automating

- Mapping automated tests to manual test cases

- Listing any bugs identified through test automation

- Listing deferred test cases for automation

Let us now understand the importance of test selection.

Creating tests for the sake of automation

I have witnessed a number of projects in which engineers have automated tests just because it is easy to automate them. This ties back to our discussion in the previous section about selecting the right tests but it deserves a special mention, as it could easily send the test automation suite into a downward spiral. The most important consideration here is to add tests that are likely to catch bugs. Adding tests just to increase the test coverage renders the suite inextensible and unmaintainable. It creates more work for the future, where tests have to be trimmed down for various reasons.

This brings us to the end of our survey on process-oriented test automation anti-patterns. Let us quickly summarize what we have learned in this chapter.

Summary

Our focus in this chapter was to understand the common issues and anti-patterns in test automation. We started by looking at how unrealistic expectations affect test automation and also how not selecting the right things to automate can impact test maintenance. We also reviewed the significance of investing in test environments and taking a whole-team and lean approach to test automation. We concluded the section by understanding the need for planning around test data. Next, we reviewed a few coding- and process-oriented anti-patterns in test automation. Some of the notable coding and design anti-patterns are compromising the code quality in tests, code duplication, incorrect use of assertions, and inefficient code reviews. Subsequently, we reviewed process-oriented anti-patterns such as test automation efforts not being assessed, test automation commencing late, and so on. This brings us to the end of our joint exploration of test automation. We have come a long way from learning about the basics of test automation to reviewing design anti-patterns in test automation frameworks. As you may have noticed, test automation involves putting together several moving parts of a complex software application. The knowledge about tools and techniques acquired throughout this book should act as a stepping stone for readers to explore test automation further.

Some potential next steps would be to dive deeper into each of the tools that we looked at and think about how you can apply them in your current role and organization. The more hands-on experience you gain with these tools and techniques, the more you will be ready to address new challenges in test automation.

Happy exploring!

Questions

1. How can test engineers and SDETs manage unrealistic expectations of test automation?

2. What is the importance of test framework maintenance?

3. How do we select the right things to automate?

4. How does code duplication affect a test framework and how do we address it?

5. What are some of the correct ways to use assertions?

6. What are some items to look for in a test automation code review?

7. What are some tasks to keep track of in test automation?

Appendix A:
Mocking API Calls

API mocking is a process in which a mock server is configured to return the response of an API with custom data. API mocking plays a principal role in test automation development. It unblocks testing in situations where an external service might not be available or is too expensive to use for testing. Mocking also helps in cases where one of the APIs is still under development and there is a need to test a chain of API calls. A classic example where API mocking helps would be to unblock frontend testing by mocking certain unavailable backend API calls. Some of the most common API mocking tools and libraries include **Jest**, **Postman**, **Cypress**, **JSON Server**, and **mocki.io**.

In this appendix, we will be looking at the following:

- How API mocking works
- Mocking API calls using Postman
- Considerations for API mocking

How API mocking works

API mocking involves creating a mock API server that returns the pre-configured response for any request from the client on that route. This mock API server is the placeholder for the actual server, which is not available for a variety of reasons. *Figure A.1* illustrates how API mocking works.

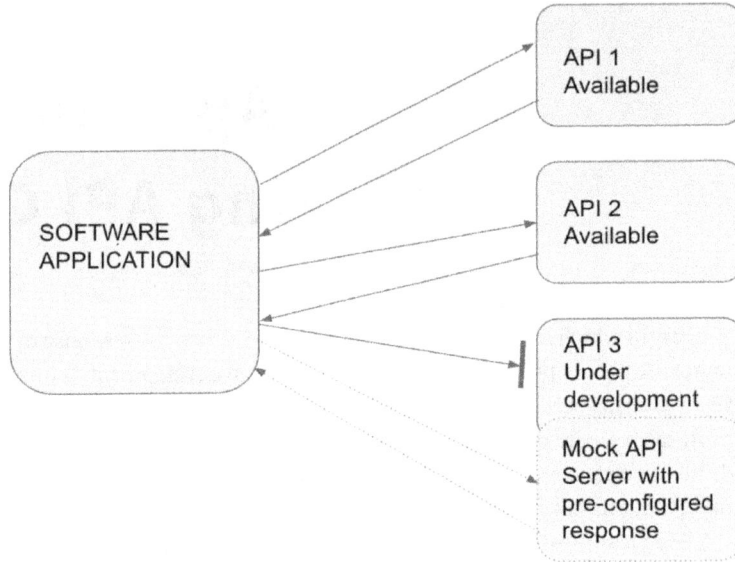

Figure A.1 – API mocking

The mock API server can be local or public and it helps temporarily remove the dependency between the frontend and backend components. It is vital to remember that the API server responds purely based on the incoming request and in no way provides a real picture of the data on the server side.

Having understood how API mocking works, let us quickly set up a mock API using Postman.

Mocking API calls using Postman

Postman provides an interactive GUI to set up mocks for API calls. Let us now review how to set one up step by step:

1. Create or use an existing Postman collection as seen earlier in *Chapter 7, Test Automation for APIs*.

2. Set up a request as shown in *Figure A.2*. There are two scenarios where we end up mocking an API request:

    I.   The first is when we have a sample response from the API call, but subsequent requests cannot be made to the API. We could save the response as an example in Postman and use it for mocking.

    II. The other one is when the API call does not exist or we do not have a sample. In this case, we will have to build the response from the scratch. We will be simulating this scenario in our example:

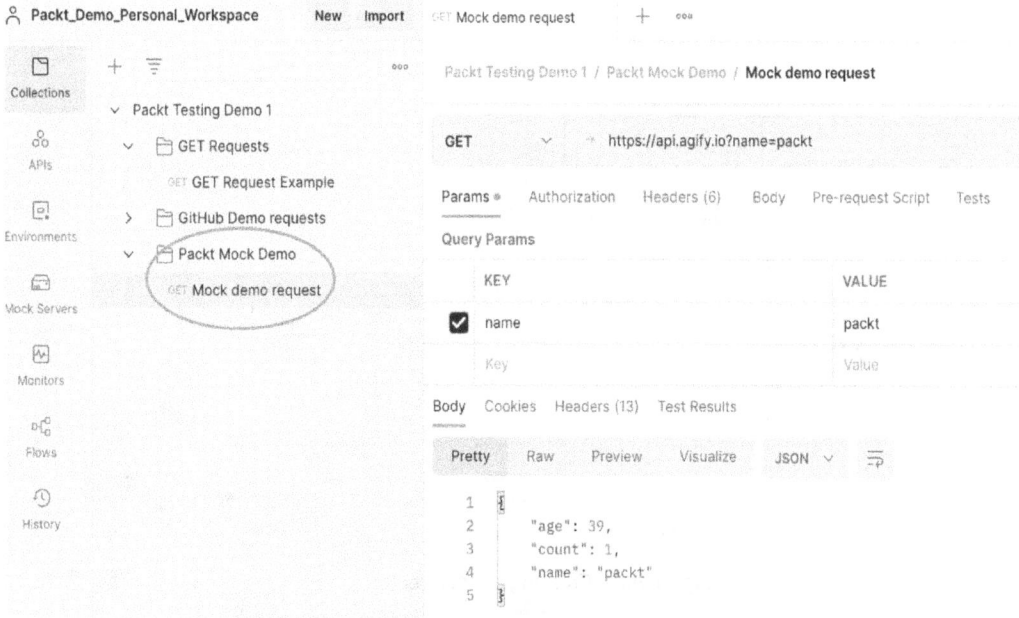

Figure A.2 – Postman request

3. The next step is to create the mock server for our request as shown in *Figure A.3*. This can be done by using the **Mock Servers** option on the left-hand pane and selecting the collection to associate it with. Once this is done, a mock server URL is generated, to which a request can be made. Postman also provides an option to make the mock server private by generating an API key to authorize requests.

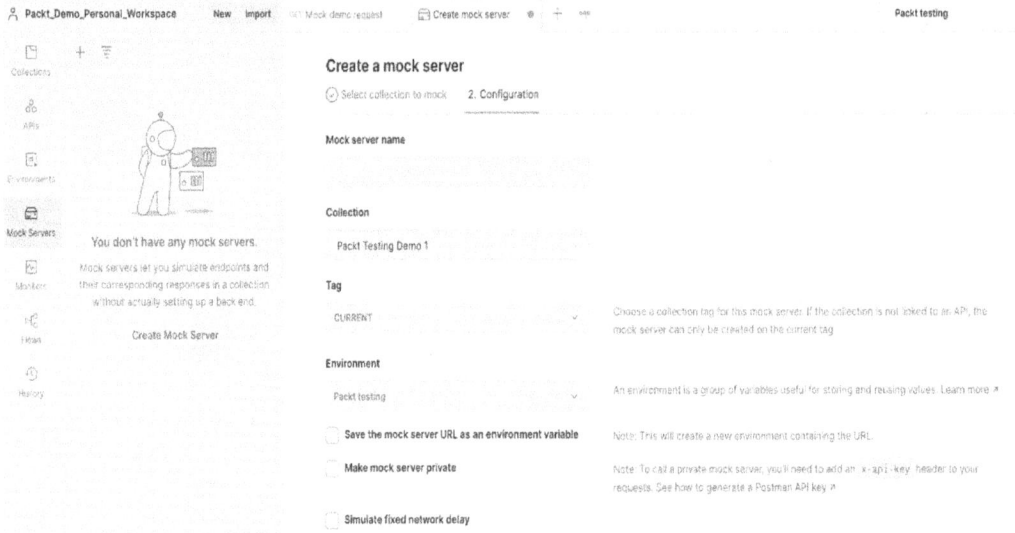

Figure A.3 – Creating a mock server

4. The next step is to add an example to the request using the **Options** menu. We will be using the dynamic variables in Postman to generate random values in our response. This can be a static response as well. We will be using the following response in our example:

```
{       "age": {{$randomInt}},
            "count": {{$randomInt}},
            "name": "{{$randomLastName}}"
}
```

5. The final step is to create a new request and use the mock server URL to make the API call. *Figure A.4* shows this in action.

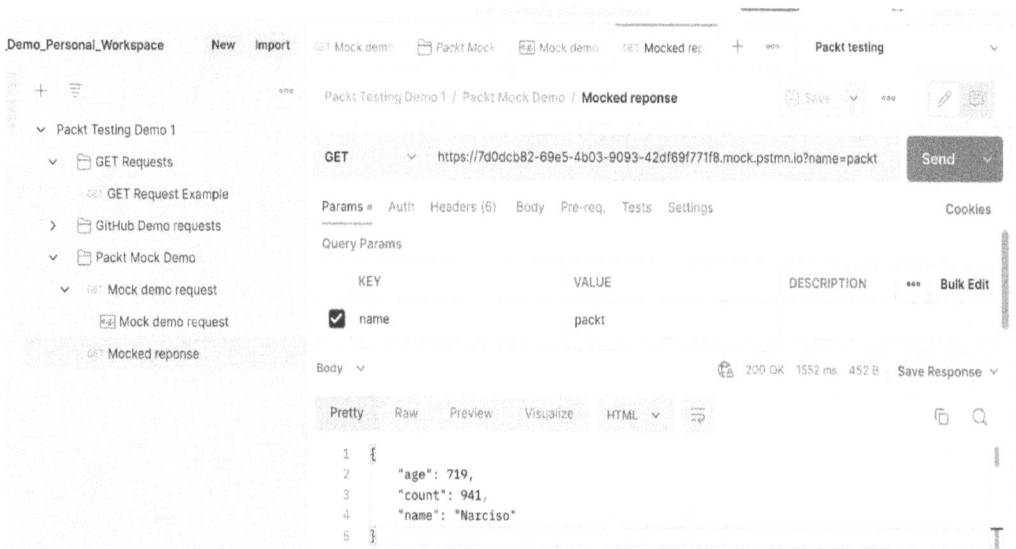

Figure A.4 – Mocked response

By using a combination of environment and dynamic variables, it is possible to simulate almost any kind of API response in Postman. Let us now look at a few important considerations when employing mocks.

Considerations for API mocking

Let us quickly review a few important considerations when mocking API calls:

- A mock API server should mimic the real responses realistically. It should conform to the same set of contracts as the real service.

- A mock API can be figured to simulate non-functional behavior such as slow response times, various error responses, and so on.

- Mock APIs usually do not persist data like real APIs and the automation test frameworks should account for their unique behaviors.

- Extra caution should be used when mocking external vendor APIs, as any uncommunicated changes on the vendor side might break the mock and bring the testing to halt.

- The transition from a mock API to a real API should be planned in advance to ensure test continuity.

Summary

This appendix helped us understand the importance of mocks in testing and how to create one quickly using Postman.

Assessments

This section contains answers to questions from all chapters.

Chapter 1, Introduction to Test Automation

1. Testing is a process to evaluate the software application with the intent to find bugs and validate product specifications.

2. Some of the key deliverables while doing software testing are the test plan, test strategy, test cases, test data, and automated test scripts.

3. Test automation is the process of setting up and validating functional and non-functional specifications using a test script. This helps increase the efficiency of testing and helps deliver value to the customer quickly.

4. Changing requirements, a lack of team collaboration, not enough planning, and a shortage of skilled resources are some of the most common challenges in the testing and test automation world.

5. A quality analyst (manual QA), test automation engineer (skilled at manual and automated testing), and an SDET are common roles in the quality engineering space.

Chapter 2, Test Automation Strategy

1. A test automation strategy not only aids in planning and executing the automated tests but also ensures that by and large, all efforts in test automation are placed in the right areas to deliver business value quickly and efficiently.

2. These are some of the main objectives of test automation:

   * Strengthening product quality

   * Improving test coverage

   * Reducing manual testing

   * Increasing test portability and reducing maintenance

   * Reducing quality costs

   * Increasing stability and reliability of the product

3. The test pyramid acts as a guideline when planning for the test coverage of a software product, and it also helps to keep in mind the scope of each test so that it remains within the designed level.

4. Cloud-based tools enable geographically dispersed teams to scale test environments up/down at will and also help in providing on-demand capacity for non-functional tests. Testing in the cloud is one of the foundational pillars of DevOps processes.

5. Some common advantages of using design patterns in test automation are set out here:

   - Helps with structuring code consistently

   - Improves code collaboration

   - Promotes reusability of code

   - Saves time and effort in addressing common test automation design challenges

   - Reduces code maintenance costs

Chapter 3, Common Tools and Frameworks

1. The **command-line interface (CLI)** is a means to interact with the shell of the system under test. A lot of the tasks performed through the graphical user interface can be done through the CLI too. But the real might of the CLI lies in its ability to programmatically support the simulation of these tasks.

2. Git is a modern distributed version control system that allows tracking changes to the source code. It primarily helps in synchronizing contributions to the source code by various members of the team by keeping track of progress over time.

3. Selenium is an open source tool that provides an interface to automate user actions on web browsers. Its main components are Selenium WebDriver, Selenium Grid, and Selenium IDE.

4. Appium is an open source tool used for automating tests on platforms such as Android and iOS. It also supports mobile browser automation with extensive community support. Appium provides libraries backing a variety of programming languages such as Java, Python, PHP, C#, JavaScript, and Ruby.

5. Cypress primarily differs from Selenium in its in-built ability to access both frontend and backend parts of the application. This makes E2E testing its focus, while Selenium specializes in browser automation.

6. JMeter is an open source performance testing tool that simulates user requests to a server and computes and stores the response statistics from the server.

7. AXE comes as a browser extension that checks the web page against pre-defined accessibility rules and generates a report of compliances/violations.

8. The criteria to be considered when selecting a test automation tool include but are not limited to the following:

   • The cost of the test automation tool and the ROI it offers over time

   • How well it fits with the team's skillset

   • CI/CD Integration capabilities

   • Support for cross-device and cross-platform test setup and execution

   • Support for custom- scripting to extend the existing framework

   • Support for various technologies such as REST, SOAP, messaging systems (Kafka), and databases

Chapter 4, Getting Started with the Basics

1. GitHub integration, debugging capabilities, build tools integration, plugins support, cost, and speed are important factors to consider when choosing an IDE.

2. `let` allows the variable to be assigned a different value later in the program but `const` does not.

3. The `indexOf` method returns the index of the element in the array. It returns `-1` if the element is not found in the array.

4. It has various advantages such as assigning multiple variables in a single statement, accessing properties from nested objects, and assigning a default value when a property doesn't exist.

5. A function is declared using the `function` keyword, followed by its name, parameters, and body. It can also have a return value.

6. This command helps identify a certain commit that is buggy or caused certain tests to fail. It uses binary search to narrow down the search to a single commit from hundreds of commits.

7. Start with a pull from remote or other branches before beginning any new work on the local code base. This keeps the local branch updated, thereby reducing merge conflicts

8. Commit messages assist fellow engineers in gaining context on why a certain change was done.

Chapter 5, Test Automation for Web

1. Cypress comes bundled with a lot of capabilities suitable for web test automation. Cypress also runs inside the browser, which allows it to test client functions directly.

2. `package.json` is the primary configuration file for npm and can be found in the root directory of the project. It helps to run your application and handle all dependencies.

3. Callback functions are mainly used to handle asynchronous aspects of modern web applications. They provide an avenue to wait and execute functions while returning control to the main block of execution.

4. `beforeEach` blocks are used within a Cypress spec to enclose code that needs to run before every `it` block.

5. The `id` attribute is selected using the # symbol.

6. `cy.intercept()` is the command to intercept and mock API calls in Cypress.

Chapter 6, Test Automation for Mobile

1. Appium involves multiple installations working in tandem, and it would be convenient to have a tool that could provide us with constant feedback on the health of our setup. `appium-doctor` does just that.

2. The main downside of using an emulator is in replicating certain performance issues such as battery life and network connectivity. The test results in these areas may be unreliable. It also gets increasingly complicated to test device behavior when multi-tasking

3. One of the major challenges is having good coverage for multiple device models and OS versions. Another challenge would be to accommodate localization testing where multiple geographical regions have to be considered for the same release of the application.

4. The benefits of using CSPs for mobile test automation are as following:

    *   Ability to scale

    *   Test and debug issues faster

    *   Save resources in setting up an in-house mobile testing lab

5. Hooks are reusable code that can be run multiple times but defined only once. Hooks help reduce test dependency, which in turn helps reduce code duplication.

Chapter 7, Test Automation for APIs

1. The primary focus of API testing is to validate the business logic of APIs, while other types of testing include performance and security. It involves the collection of test data to be sent in as JSON requests to the API and validating the response for accuracy. Some items to look for in the response are status, syntax, schema conformance, and functional correctness.

2. Postman provides a way to group API requests using collections. This helps organize a workspace by breaking it down, and a workspace can also be sorted into multiple collections.

3. Snippets are small code blocks readily available in Postman and can be selected for use in tests. They give an overview of frequently used API validations and save time in writing them.

4. Postman allows the chaining of API requests, which enables the usage of variables from one request to another. In this manner, multiple requests can be tested as a series of calls.

5. Newman is a command-line tool used in conjunction with Postman to execute collections.

6. Docker is a platform that assists in building, deploying, and testing your application code on units called containers, irrespective of the underlying operating system. It provides great portability for developing and testing applications.

7. GraphQL is a query language specification for APIs and helps design APIs in a client-driven architecture. It excels at getting all the data needed by the client in a single request.

Chapter 8, Test Automation for Performance

1. JMeter is a performance-testing application built using Java. It is a completely free and open source tool created by Apache and it can be used to performance test a wide variety of applications, including APIs and databases.

2. JMeter uses the thread group component to simulate user load by specifying the number of threads, the ramp-up period, and the loop count.

3. Performance tests are often long running and tend to be heavy on system resource consumption. GUI mode consumes a lot of memory, especially when running pre-recorded scripts, and execution via the command line alleviates this pain by reducing the memory footprint of the tool. Additionally, the CLI offers integration with CI/CD.

4. There may be instances where the features that come out of the box with JMeter are not sufficient and custom scripts are needed to perform specific tasks. JSR233 and Beanshell assertions/samplers can be utilized in cases such as these to get the job done.

5. Distributed load testing is the process of mimicking the real-world user load and distributing the load across multiple machines/servers to uncover critical bugs or memory leaks.

Chapter 9, CI/CD and Test Automation

1. CI/CD stands for Continuous Integration and Continuous Delivery. These software engineering practices allow us to automate the building and testing of code to validate it meets the specifications. After all the tests have passed, they equip the teams with options to automate the delivery of code. Continuous Integration, Continuous Delivery, and Continuous Deployment together significantly reduce the time it takes to deliver application enhancements while the engineering team can solely focus on product and code enhancements.

2. A CI process is triggered when an engineer pushes code changes to the version control repository. A new instance of the CI pipeline is triggered where build and tests are executed.

3. A CI server typically keeps polling the version control repository every minute or so and runs a preconfigured build script by fetching the latest version of the source files.

4. Some of the key aspects involved in the CD methodology are:

- Effective software configuration management

- CI

- Building and maintaining a deployment pipeline

- Automatic infrastructure and environment management

5. E2E tests should be configured to run on the pipeline for every deployment to a new environment.

6. GitHub Actions is a CI/CD platform that enables the automation of building, testing, and deployment of application code. It is the built-in CI/CD tool for GitHub.

Chapter 10, Common Issues and Pitfalls

1. It is the duty of SDETs and quality engineers to educate the stakeholders about the initial investment it takes see results from test automation. They should also highlight the importance of having skilled resources to address the complex application stack.

2. Working on any kind of design work without considering maintenance aspects does not scale very well. In the pilot stages of a software project, it is vital to think through what kind of effort would go into maintaining a test automation framework after it is fully built and functional.

3. Some ideas to select the right things to automate are as following:

- Do not automate tests that do not need to be run frequently

- Focus on stable business scenarios that can reap the benefit in the long run

- Minimize automating tests at the UI layer Do not automate the usability aspects of your application

- Avoid automating big and complex scenarios

4. Code duplication hurts the maintainability of automation frameworks overall. It is important to remember the **Rule of three** when it comes to code duplication. Whenever there is a need to duplicate a chunk of code the third time, it needs to be abstracted into a separate method or class.

5. A few key ways to use assertions are:

- Certainly make use of assertions for validations in your test

- Use multiple assertions in a test sparingly and only where necessary

- Use appropriate type of assertions where applicable. Do not use the same kind of assertion for all validations

6. Some important aspects to look for when reviewing test automation code are:

- Formatting and Typos

- Test execution failures

- Hard-coded and static handling of variables

- Potential for test flakiness

- Design pattern deviations

- Code duplication

- Any invalid structural changes to the framework

- Correct handling of config files and their values

7. Some critical items to keep a track of when it comes to test automation are:

- What is automated/not automated?

- Tests not worth automating

- Mapping of automated tests to manual test cases

- List of bugs found identified through test automation

- List of deferred test cases for automation

Index

A

A/B testing 17

acceptance testing 17

Agile environment
  test automation 11
  testing, demands 5

Agile methodology 17

Agile test automation
  quadrants 27
  strategy, implementing 26-28

Android emulator
  configuring 129-132

anti-pattern 217

API automation
  considerations 171, 172

API calls
  intercepting 119, 120
  mocking, considerations 233

API mocking 229
  working 229, 230

API requests
  chaining 166, 167
  organizing with collections 157-161

API response
  asserting, with snippets 162-164

API testing 18

API testing tool
  selecting, criteria 67

API tests 205

Appium 60, 126
  advantages 126
  components 60
  high-level architecture 60, 61
  installing 127, 128
  reference link 61, 141
  WebdriverIO, configuring with 132, 133

Appium Client 60

Appium doctor 142

Appium Inspector
  configuring 136-138
  installing 136-138

Appium Server 60

Appium/WebdriverIO test 140, 141

application programming interfaces
    (APIs) 10, 34, 147

arrow functions 109
  creating, in JavaScript 109

assertions
  working with 184, 185

async/await functions
  using, in JavaScript 138, 139

automated API tests
  writing 162
automation engineer
  tools 43
Axe 65, 66
  reference link 65

B

Bamboo 32
Bash
  reference link 48
Behavior-Driven Development (BDD) 18
black-box testing 18
bug
  versus defect 6
business layer pattern 41

C

callback functions 110
  creating, in JavaScript 110, 111
Carthage 142
Chai 112
Chai Assertion Library
  reference link 162
Chef 25
CI/CD integration 68
CircleCI 104
CLI 44-46
  advantages 48
  flags 47, 48
  shell scripting 48, 49
  usage, tips/tricks 47
  Vim, working with 46, 47
cloud-based environment
  tapping, advantages 34

cloud service providers (CSPs) 143
coding/design anti-patterns
  assertions, incorrect usage 220, 221
  business logic, validating with UI tests 222
  code duplication, impact 218, 219
  code quality, compromising in tests 217
  complex tests, affecting test
    maintainability 219, 220
  data, handling within test
    frameworks 221, 222
  inefficient code reviews, effects 223
  refactoring, significance 222
  tightly coupled code 218
cohesion 218
collections
  used for organizing API requests 157-161
collection variables 164
command line
  used, for executing spec 113-115
  used, for handling Jmeter's tests 186-188
comma-separated values (CSV) file 30
component tests 205
conditional statements
  using 97-99
containers 171
continuous delivery (CD)
  and deployment pipeline 202-204
**continuous integration and continuous
  delivery (CI/CD) 199**
  basics 201, 202
  GitHub Actions 208-212
  process 200
  test automation strategies 204
continuous integration (CI) 32, 104
continuous testing 10
coupling 218
Create, Read, Update, Delete (CRUD) 148

cy.get function 116
Cypress 62, 104, 229
    additional configurations 120, 121
    API calls, intercepting 119, 120
    components 62
    features 104
    high-level architecture 62, 63
    installing 104-109
    limitations 121
    reference link 63
    selectors, using 116
    test, creating 109
Cypress Node.js server 62
Cypress test runner 62

D

data-driven testing 18
data types
    objects 87
    primitive 87
data variables 164
defect 6
    versus bug 6
defect management
    in testing 6
design patterns, in test automation
    business layer pattern 41
    factories pattern 40, 41
    Page Object Model (POM) 38-40
    using 37
development-operations (DevOps) 7, 34
Docker
    URL 148
Document Object Model (DOM) 116
dynamic test suite data 221

E

ECMAScript 84
ECMAScript 6 (ES6) version 109
emulator 143
**end-to-end (E2E) test automation
    framework** 127
end-to-end (E2E) testing 18, 25, 104
end-to-end (E2E) tests 200, 205, 206
environment variables 164
exploratory testing 18
Extreme Programming 9

F

factories pattern 40, 41
fat arrow (=>) 109
flags
    in CLI 47, 48
flags, using with commit command
    --amend 75-77
    -m 74, 75
framework level constants 222
functional bugs 4
functional test data 221

G

Garden.io 216
GET API request
    creating 152-155
GET request
    sending with Postman 151
Git 49-56, 74
    change, committing 74
    commands 80

commit message 74

flags 76

flags, using with commit command 74-77

merge conflicts, resolving 77-80

reference link 56

Git commands

git bisect 80

git cherry-pick 80

git rebase 81

git reflog 81

git reset 81

git revert 81

GitHub Actions 104

global data 221

global variables 164

go-to-market (GTM) 34

Grid 58

Groovy documentation

reference link 193

H

Homebrew 85, 87

HTTP(S) Test Script Recorder

using 189, 190

Hybrid apps 60

I

**Integrated Development
Environment (IDE) 58**

selecting, features 81, 82

setting up 82, 83

using 81

VS Code, downloading 82, 83

integration testing 18

Internet of Things (IoT) 29

**iOS-specific considerations,
mobile automation**

CSPs, selecting 143

real devices, versus emulators 143

issues

test automation 213-217

J

Java 191-193

Java Development Kit (JDK) 193

Java documentation

reference link 193

Java Runtime Environment (JRE) 193

JavaScript 84

basics 85

data types 87, 88

exploring 100

functions 99, 100

learning, need for 84

objects 89

strings, working with 88, 89

variables 86, 87

JavaScript arrays

using 89-91

JavaScript functions

with async/await 138, 139

JavaScript objects

arrays 94

destructuring 93

object literals, working with 91, 92

JavaScript program

executing 85

Node.js, installing 84

running 84

Java Virtual Machine (JVM) 191, 193

Jest 229

JIRA 32

JMeter 63, 176

components 64

download link 176

high-level architecture 64, 65

HTTP(S) Test Script Recorder,
    using 189, 190

installing 176-178

Java 191-193

Java essentials for 190

performance test, automating 179

performance testing, considerations 195

reference link 64

working 176

JMeter master 64

JMeter slave nodes 64

JMeter, user manual

reference link 190

JSON Server 229

JSR233 assertion

using 193-195

L

latency

versus sample time 184

listener 182

load testing 18

local variables 164

loops 94

working with 95, 96

M

machine learning (ML) 28

master on repository 49

master-slave architecture 64

mobile automation 141

challenges 141, 142

iOS-specific considerations 142

mobile automation framework

optimizing 144-146

mobile test

writing 138

mobile testing tools 68

Mocha framework 112, 133

mock API

setting up, with Postman 230-233

Mocki.io 229

N

Native apps 60

Newman 169

Node.js 60, 84

non-functional requirements (NFRs) 204

O

object-oriented programming (OOP) 104

P

Page Object Model (POM) 38-40

penetration testing 18

performance testing tool

requirements 66

selecting 66

performance test, JMeter
  assertions, working with 184, 185
  automating 179
  building 179-184
  running 179-184
  working, via command line 186-188
**personally identifiable information
      (PII) 33, 122**
Portable Document Format (PDF) file 32
POST API request
  creating 155-157
Postman 229
  API requests organizing with
      collections 157-161
  automated API tests, writing 162
  download link 149
  REST API testing 148
  tests, executing 167-171
  used for sending GET request 151
  used for sending POST request 151
  using, to set up mock API 230-233
  variables 164-166
  working with 148
  workspaces, creating 150
  workspaces, managing 150
Postman application
  downloading 149
Postman docs
  reference link 171
POST request
  sending with Postman 151
Prettier 112
process-oriented anti-patterns 223-226
Progressive Web Apps (PWAs) 142

Q

quality 7
quality engineering
  roles, exploring 15
Quick Test Professional (QTP) 30

R

real devices 143
request dashboard in Postman 151
REST APIs
  testing aspects, GraphQL versus 172
REST API testing 148

S

sample time
  versus latency 184
security testing 18
selectors 116
  asserting on 117-119
  working with 116, 117
Selenium 57
  components 58
  high-level architecture 58, 59
Selenium WebDriver 40
shell scripting 48
Shift-Left approach 7
Shift-Right approach 7
smoke testing 18
smoke tests 206, 207
snippets
  used, for asserting API response 162-164

Software Development Engineer in Test (SDET) 16, 17, 32 , 204
  responsibilities 16
  versus test automation engineer 16
software testing 4
spec 109
  executing 113
  executing, with command line 113-115
  executing, with visual test runner 115, 116
  structure 112
  writing 112
stress testing 18
system testing 18

T

Terraform 25
test automation 9-11
  agile test automation 11, 12
  anti-patterns 217
  challenges 12
  coding/design anti-patterns 217
  metrics 13, 14
  process-oriented anti-patterns 223-226
  recurrent issues 213-217
  regression bugs, finding 13
  regression bugs, handling 13
test automation engineers 15, 16
test automation environment 24
  constituting 24, 25
  provisioning 25
  testing, in production environment 25
test automation frameworks 56
  Appium 60
  Axe 65
  Cypress 62
  JMeter 63

  selecting 66
  Selenium 57
  standards 30, 31
  types 30, 31
test automation framework, standards
  maintaining 32, 33
  platform, laying for tests 32
  test data, sanitizing 33, 34
  test libraries, managing 31
test automation strategies, for CI/CD
  API tests 205
  E2E tests (API and UI) 205, 206
  smoke tests 206, 207
  test failures, addressing 207
  unit/component tests 205
test automation strategy 22
  devising 29
  management support, obtaining 23
  objectives 22, 23
  scope, defining 24
  testing, in cloud 34
  test results, reporting 28, 29
  tools and training, selecting 29
test automation tool, selecting considerations 68
  budget considerations 68
  CI/CD integration 68
  team's skillset 69
test case 18
test, Cypress 109
  arrow functions, creating in JavaScript 109, 110
  callback functions, creating in JavaScript 110, 111
  creating 109
  spec, executing 113

spec structure 112

writing 112

Test-Driven-Design (TDD) 9

Test-Driven Development (TDD) 19

test early and test often 8, 9

test failures

addressing 207, 208

testing 3

challenges 8

defect management 6

demands, in agile environment 5

DevOps 7

functional 4

need for 4

non-functional 4

quality 7

Shift-Left approach 7

Shift-Right approach 7

tasks 4, 5

testing aspects, GraphQL

versus RestAPI 172

test plan 19, 176

test pyramid 35

API tests 36

component tests 35, 36

E2E tests 36

integration tests 36

system tests 36

test cycles, structuring 37

tests 36

UI tests 36

unit tests 35, 36

tests

executing, in Postman 167-171

test suite/test automation suite 19

time-to-market (TTM) 34

tools

for automation engineer 43

selecting 66

TypeScript 113

U

UIAutomator 126

Unified Functional Testing (UFT) tool 30

unit tests 205

unit testing 19

usability testing 19

user acceptance testing (UAT) 25

user interface 64

V

validation 19

verification 19

Vim

reference link 46

working with 46, 47

virtual machines (VMs) 24

visual test runner

used, for executing spec 115, 116

W

waterfall model 5, 19

Web apps 60

web automation

considerations 121, 122

WebDriver 58

WebdriverIO 127

advantages 127

Android configuration 134-136

configuring, with Appium 132, 133

web testing tool
features 67
selecting 67
white-box testing 19

X

Xcode 142
Xcode Command Line Tools 142
XCUITest 126, 142

Packt>

Other Books You May Enjoy

If you enjoyed this book, you may be interested in these other books by Packt:

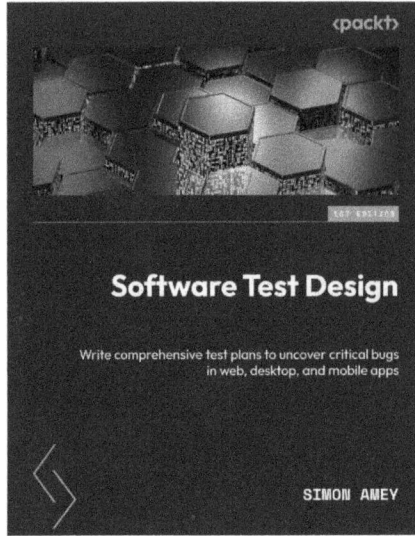

Software Test Design

Simon Amey

ISBN: 978-1-80461-256-9

- Understand how to investigate new features using exploratory testing
- Discover how to write clear, detailed feature specifications
- Explore systematic test techniques such as equivalence partitioning
- Understand the strengths and weaknesses of black- and white-box testing
- Recognize the importance of security, usability, and maintainability testing
- Verify application resilience by running destructive tests
- Run load and stress tests to measure system performance

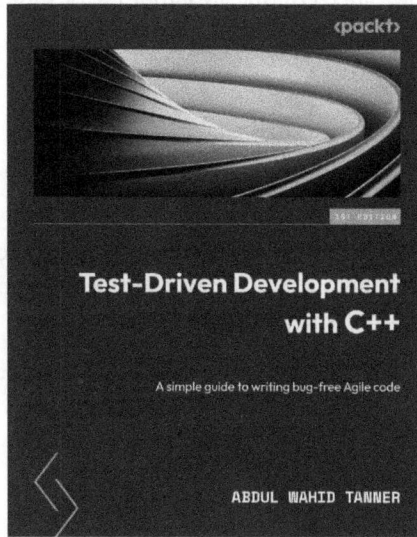

Test-Driven Development with C++

Abdul Wahid Tanner

ISBN: 978-1-80324-200-2

- Understand how to develop software using TDD
- Keep the code for the system as error-free as possible
- Refactor and redesign code confidently
- Communicate the requirements and behaviors of the code with your team
- Understand the differences between unit tests and integration tests
- Use TDD to create a minimal viable testing framework

Packt is searching for authors like you

If you're interested in becoming an author for Packt, please visit `authors.packtpub.com` and apply today. We have worked with thousands of developers and tech professionals, just like you, to help them share their insight with the global tech community. You can make a general application, apply for a specific hot topic that we are recruiting an author for, or submit your own idea.

Share Your Thoughts

Now you've finished *Test Automation Engineering Handbook*, we'd love to hear your thoughts! Scan the QR code below to go straight to the Amazon review page for this book and share your feedback or leave a review on the site that you purchased it from.

`https://packt.link/r/1804615498`

Your review is important to us and the tech community and will help us make sure we're delivering excellent quality content.

Download a free PDF copy of this book

Thanks for purchasing this book!

Do you like to read on the go but are unable to carry your print books everywhere?

Is your eBook purchase not compatible with the device of your choice?

Don't worry, now with every Packt book you get a DRM-free PDF version of that book at no cost.

Read anywhere, any place, on any device. Search, copy, and paste code from your favorite technical books directly into your application.

The perks don't stop there, you can get exclusive access to discounts, newsletters, and great free content in your inbox daily

Follow these simple steps to get the benefits:

1. Scan the QR code or visit the link below

https://packt.link/free-ebook/9781804615492

2. Submit your proof of purchase
3. That's it! We'll send your free PDF and other benefits to your email directly

www.ingramcontent.com/pod-product-compliance
Lightning Source LLC
Chambersburg PA
CBHW080521220326

41599CB00032B/6165

* 9 7 8 1 8 0 4 6 1 5 4 9 2 *